P9-DWP-898

93

# Advances in Biochemical Engineering/Biotechnology

Series Editor: T. Scheper

# Advances in Biochemical Engineering/Biotechnology

**Series Editor: T. Scheper**

Recently Published and Forthcoming Volumes

**Microscopic Techniques**
Volume Editor: Rietdorf, J.
Vol. 95, 2005

**Regenerative Medicine I**
**Clinical and Preclinical Applications**
Volume Editor: Yannas, I.V.
Vol. 94, 2005

**Regenerative Medicine II**
**Theories, Models and Methods**
Volume Editor: Yannas, I.V.
Vol. 93, 2005

**Technology Transfer in Biotechnology**
Volume Editor: Kragl, U.
Vol. 92, 2005

**Recent Progress of Biochemical and Biomedical Engineering in Japan II**
Volume Editor: Kobayashi, T.
Vol. 91, 2004

**Recent Progress of Biochemical and Biomedical Engineering in Japan I**
Volume Editor: Kobayashi, T.
Vol. 90, 2004

**Physiological Stress Responses in Bioprocesses**
Volume Editor: Enfors, S.-O.
Vol. 89, 2004

**Molecular Biotechnology of Fungal $\beta$-Lactam Antibiotics and Related Peptide Synthetases**
Volume Editor: Brakhage, A.
Vol. 88, 2004

**Biomanufacturing**
Volume Editor: Zhong, J.-J.
Vol. 87, 2004

**New Trends and Developments in Biochemical Engineering**
Vol. 86, 2004

**Biotechnology in India II**
Volume Editors: Ghose, T.K., Ghosh, P.
Vol. 85, 2003

**Biotechnology in India I**
Volume Editors: Ghose, T.K., Ghosh, P.
Vol. 84, 2003

**Proteomics of Microorganisms**
Volume Editors: Hecker, M., Müllner, S.
Vol. 83, 2003

**Biomethanation II**
Volume Editor: Ahring, B.K.
Vol. 82, 2003

**Biomethanation I**
Volume Editor: Ahring, B.K.
Vol. 81, 2003

**Process Integration in Biochemical Engineering**
Volume Editors: von Stockar, U., van der Wielen, L.A.M.
Vol. 80, 2003

**Microbial Production of L-Amino Acids**
Volume Editors: Faurie, R., Thommel J.
Vol. 79, 2003

**Phytoremediation**
Volume Editor: Tsao, D.T.
Vol. 78, 2003

**Chip Technology**
Volume Editor: Hoheisel, J.
Vol. 77, 2002

**Modern Advances in Chromatography**
Volume Editor: Freitag, R.
Vol. 76, 2002

**History and Trends in Bioprocessing and Biotransformation**
Vol. 75, 2002

# Regenerative Medicine I
# Theories, Models and Methods

Volume Editor: Ioannis V. Yannas

With contributions by
C. E. Butler · M. K. Call · A. S. Colwell · M. DeFrances
M. T. Longaker · H. P. Lorenz · A. L. Mescher · G. K. Michalopoulos
A. W. Neff · D. P. Orgill · D. L. Stocum · P. A. Tsonis · I. V. Yannas

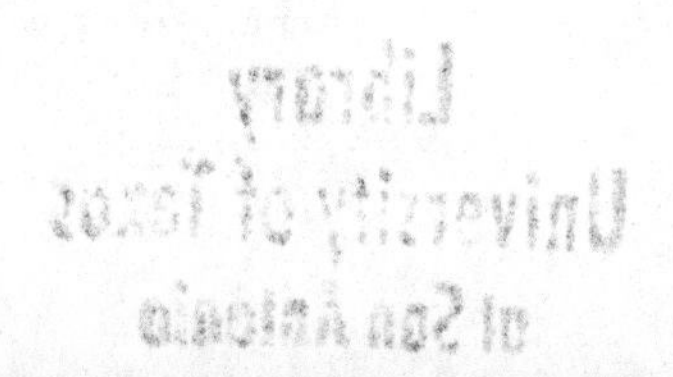

*Advances in Biochemical Engineering/Biotechnology* reviews actual trends in modern biotechnology. Its aim is to cover all aspects of this interdisciplinary technology where knowledge, methods and expertise are required for chemistry, biochemistry, micro-biology, genetics, chemical engineering and computer science. Special volumes are dedicated to selected topics which focus on new biotechnological products and new processes for their synthesis and purification. They give the state-of-the-art of a topic in a comprehensive way thus being a valuable source for the next 3–5 years. It also discusses new discoveries and applications.

Special volumes are edited by well known guest editors who invite reputed authors for the review articles in their volume.

In references *Advances in Biochemical Engineering/Biotechnology* is abbreviated as *Adv Biochem Engin/Biotechnol* as a journal.

Visit the ABE home page at springeronline.com

Library of Congress Control Number 2004110174

ISSN 0724-6145
ISBN 3-540-22871-3 **Springer Berlin Heidelberg New York**
DOI 10.1007/b14095



**Springer is a part of Springer Science+Business Media**
springeronline.com

Printed in The Netherlands



Typesetting: Fotosatz-Service Köhler GmbH, Würzburg
Cover: KünkelLopka GmbH, Heidelberg; design & production GmbH, Heidelberg

Printed on acid-free paper 02/3141 – 5 4 3 2 1 0

## Series Editor

Professor Dr. T. Scheper
Institute of Technical Chemistry
University of Hannover
Callinstraße 3
30167 Hannover, Germany
*scheper@iftc.uni-hannover.de*

## Volume Editor

Professor Dr. Ioannis V. Yannas
Massachusetts Institute of Technology
77 Massachusetts Avenue
Cambridge, MA 02139-4307
USA
*yannas@mit.edu*

## Editorial Board

## Advances in Biochemical Engineering/Biotechnology Also Available Electronically

For all customers who have a standing order to Advances in Biochemical Engineering/Biotechnology, we offer the electronic version via SpringerLink free of charge. Please contact your librarian who can receive a password for free access to the full articles by registering at:

springerlink.com

If you do not have a subscription, you can still view the tables of contents of the volumes and the abstract of each article by going to the SpringerLink Homepage, clicking on "Browse by Online Libraries", then "Chemical Sciences", and finally choose Advances in Biochemical Engineering/Biotechnology.

You will find information about the

- Editorial Board
- Aims and Scope
- Instructions for Authors
- Sample Contribution

at springeronline.com using the search function.

## Attention all Users
## of the "Springer Handbook of Enzymes"

Information on this handbook can be found on the internet at **springeronline.com** using the search function.

A complete list of all enzyme entries either as an alphabetical Name Index or as the EC-Number Index is available at the above mentioned URL. You can download and print them free of charge.

A complete list of all synonyms (more than 25,000 entries) used for the enyzmes is available in print form (ISBN 3-540-41830-X).

**Save 15%**

We recommend a standing order for the series to ensure you automatically receive all volumes and all supplements and save 15% on the list price.

# Preface

How do you grow back a disabled organ? The key is to find a way to induce tissue and organ regeneration. This prospect is clearly different from spontaneous phenomena such as compensatory growth of the adult liver or kidney; nor is it the epimorphic regeneration of limbs that is spontaneously observed with certain amphibians. It is the induced regrowth of an organ at the anatomical site of an adult. There the function of the original organ has been lost, either following accidental trauma or elective surgery or, conceivably, after an organ has become dysfunctional due to a chronic insult.

This is a new field of study; yet, this volume provides answers from well-known investigators for each of several adult organs. These are early efforts and the regenerated organs are often quite imperfect. Nevertheless, in some cases, the clinical benefit appears to be highly significant; occasionally, even unique and indispensable.

In spite of its rough and undeveloped contours, the methodology of induced regeneration has already yielded spectacular results over the entire anatomy. There are chapters on induced regeneration of heart valves, peripheral nerves, skin, cartilage, urological organs, the conjunctiva, bone, liver, the spinal cord. Useful background is provided by a chapter on the compensatory hypertrophy of the liver, an unusual and instructive healing process in the adult. And, in addition, there are also chapters that describe other spontaneous regeneration phenomena such as those observed in the mammalian fetus or limb (epimorphic) regeneration observed amphibians. Background chapters of this type are useful as relevant underpinnings, or even possibly as "controls", for data from the adult mammal. An experimental tool, the use of stem cells, is treated separately. Two emerging theoretical approaches, the fetal healing reactivation theory and the immunocompetence theory, are also described. Methodology and theories have been selected by the editor on the expectation that each will become increasingly useful in tomorrow's investigations.

Regeneration of adult organs is an unexpected fact. It is the mammalian fetus, not the adult, who regenerates spontaneously following severe injury to an organ. Following the remarkable pioneering efforts of the 1950s and 1960s that led to transplantation of the kidney and the heart, it had become widely accepted that loss of adult organ function could only be treated by heroic efforts. A serendipitous discovery was made in the mid-1970s that changed this thinking. Studies with animals, and later with burn patients who had lost large areas of their skin, showed that the healing processes of deep skin wounds could be controlled by use of a protein scaffold, misnamed "artificial skin" at that time. Unexpectedly and dramatically, but quite confusingly for a while, this scaffold changed the adult healing process from contraction and scar formation to partial regeneration of skin. The new skin was imperfect, lacking hair and sweat glands, but it was made up of a normal epidermis and, most importantly, a normal dermis as well as a normal epidermal-dermal junction.

It is now clear that at least partial regeneration of several organs can be obtained by the use of no more than three classes of relatively simple "reactants": cell suspensions, solutions of growth factors, and insoluble matrices. The rules that specify the required reactants for "synthesis" of tissues and organs are emerging rapidly. There is evidence that regeneration of the stroma, perhaps the most challenging problem in organ regeneration, does not require the investigation to use exogenous cells or exogenous growth factors but only requires grafting at the injured site of a degradable scaffold, a "regeneration template", with a highly specific structure. In general, the methodology of organ regeneration is far simpler than that used for organ transplantation and can be largely defined, even occasionally standardized, using physicochemical and biochemical criteria. The new methodology is often referred to as tissue engineering. One successful paradigm has consisted of *de novo* organ synthesis at the correct anatomical site using cell-seeded scaffolds. In other cases, organ synthesis has been achieved *in vitro* prior to implantation at the correct anatomical site.

A fascinating possibility emerges after the data on induced organ regeneration have been examined: The fetal response to injury theoretically lies fast asleep in the adult but it can be awakened by use of appropriate tools and harnessed to grow back tissues and organs. Another interesting prospect emerges following the discovery of associations between the fetal processes of loss of regenerative potential and the concurrent acquisition of immunocompetence. These recent theoretical advances provide formidable and exciting challenges to future investigators of organ regeneration in adults.

The editor would like to thank the staff at Springer-Verlag who have helped him during the preparation of this volume. My thanks go to Professor Dr. Thomas Scheper, Institut für Technische Chemie, Hannover University, Germany as well as to Dr. Marion Hertel, Chemistry Editor, and especially to Ms. Ulrike Kreusel, Chemistry Desk Editor.

Cambridge, October 2004 Ioannis V. Yannas

# Contents

# Contents of Volume 94

## Regenerative Medicine II

### Clinical and Preclinical Applications

Volume Editor: Ioannis V. Yannas

Adv Biochem Engin/Biotechnol (2005) 93: 1–38
DOI 10.1007/b99965


# Facts and Theories of Induced Organ Regeneration

Ioannis V. Yannas (✉)

Massachusetts Institute of Technology, 77 Massachusetts Avenue, Cambridge, MA02139-4307, USA
*yannas@mit.edu*



**Abstract** Induced organ regeneration is de novo synthesis of a physiological, or nearly physiological, organ at the same anatomical site as the organ that is being replaced. Regeneration of skin, peripheral nerves and the conjunctiva, described in this chapter, have been accomplished using biologically active scaffolds (regeneration templates) seeded with epithelial cells; devices for regeneration of the first two organs are in clinical use. There is substantial empirical evidence that templates induce regeneration by blocking contraction, the major mechanism for closure of severe wounds in adults. Templates appear to function by interfering with normal myofibroblast function well as by acting as temporary configurational guides for synthesis of new stroma that resembles that of the organ under replacement. The combined evidence supports a theory which predicts that selective blocking of the adult healing response uncovers the latent fetal response to injury and leads to organ regeneration. An independent theory suggests that loss of regenerative potential during the mammalian fetal-adult transition is associated with simultaneous acquisition of individual immunocompetence.

**Keywords** Regeneration · Skin · Peripheral nerves · Myofibroblasts · Contraction

**List of Abbreviations and Symbols**

BM Basement membrane
$C$ % Wound closure by contraction
CB Contraction blocking by template
CBSIR Contraction-blocking synchronous isomorphous replacement theory
DRT Dermis regeneration template
ECM Extracellular matrix
$\mathbf{F}_c$ Macroscopic force vector of wound contraction
$f_i$ Contractile force vector per cell
GAG Glycosaminoglycan
MFB Myofibroblast(s)
MRL Mouse strain
$N$ Number of contractile cells in a wound
PDGF Platelet-derived growth factor
PNS Peripheral nervous system
$R$ % Wound closure by regeneration
$S$ % Wound closure by scar formation
SIR Synchronous isomorphous replacement of template by new tissue during regeneration
$t_d$ Time constant for template degradation
TGF-β1 Transforming growth factor-beta1
$t_h$ Time constant for tissue synthesis during healing
$\Delta$ Change in a property
$\sum f_i$ Sum of individual cell force vectors in a wound
$\rho_c$ Volume density of cells
$\sigma$ Specific surface of template
$\Phi_c$ Surface density of cells

# 1 Introduction

The adult organism responds spontaneously to a severe injury by mounting a healing response that spares the organism but condemns the injured organ. In contrast, the fetal healing response leads to regeneration of the injured organ. Induced regeneration of an adult organ is the result of an experimental manipulation that modifies drastically the adult response, making it very similar to that of the fetus. Since there is a great dearth of organs for transplantation in our hospitals there is obvious clinical benefit to procedures that lead to induced regeneration.

In contrast to processes such as physiological regeneration [1], compensatory growth [2], epimorphic regeneration [2–4], or repair [1–3], induced regeneration refers to recovery of the nonregenerative tissues of the adult mammal [5]. Organ regeneration is distinct from organ repair as an endpoint of a healing process following injury. Repair is a physiological adaptation to loss of normal organ mass and leads to restoration of the interrupted continuity by synthesis of scar tissue without recovery of the uninjured tissues [2]. In contrast, regeneration restores the interrupted continuity by synthesis of the missing organ at the original site, yielding a regenerate. Regeneration restores the normal structure and function of the organ; repair does not.

The earliest reports of this phenomenon have interchangeably used the terms "organ replacement" [6], "synthesis" [7–9] and "regeneration" [10, 11] to describe the structural and functional recovery of the organ at the original anatomical site. Induced organ regeneration is defined as de novo synthesis [5, 12] of a physiological, or nearly physiological, organ at the same anatomical site as the organ that is being replaced. Being a process of de novo synthesis it makes use of elementary reactants. These comprise cells of various types, soluble regulators (growth factors, cytokines or other soluble signaling molecules) and insoluble regulators of cell function (matrices or scaffolds). The definition of induced regeneration excludes use of transplantation or autografting of whole tissues and organs.

Historically, the vast majority of studies on induced regeneration have been conducted with just two organs, skin and peripheral nerves. Accordingly, the scope of this chapter, focused as it is on a body of related data and on their interpretation, is limited to discussion of induced regeneration of these two organs. A third organ, the conjuctiva, is also discussed in this chapter (also in Hatton and Rubin, this volume). Other chapters in this volume deal with regeneration of diverse organs. Although skin and peripheral nerves are quite different in structure and function, the similarities in response of these two organs to injury is so striking that intriguing "trans-organ" principles can be readily developed [5]. Such principles become the basis for a generalized theoretical approach that might hypothetically pertain to other organs as well.

By definition, regeneration processes occur in situ, i.e., at the precise anatomical site where the organ under replacement was located. In spite of the eventual implantation of the regenerated organ in vivo (inside the animal or human host), however, the protocol for synthesis of the organ may be carried on in cell culture on the laboratory bench (in vitro) until the product is ready to be implanted. Although the complexity of such in vitro methodology may be extensive, involving lengthy protocols for cell culture, the resulting organ-like preparation (organoid) must eventually survive its implantation. Since implantation normally takes place at a site that has been recently injured during its surgical preparation, an organoid must emerge from the in vitro stage ready to adapt to the ongoing healing processes at the implantation site. The use of cells, protein solutions and scaffolds to synthesize organs by in vitro or in vivo procedures has been often referred to as tissue engineering.

In this emerging field of investigation the organs that have so far been induced to regenerate are not perfectly physiological in structure and function. Even though only imperfectly physiological, however, the regenerate is the documented new product of the healing process that has displaced contraction and scar formation as a closure mode for the injured organ. Furthermore, at least two among these imperfect regenerates are FDA-approved devices that are used for treatment of skin loss and for paralysis of limbs due to trauma. The clinical benefit is real and highly unusual. These considerations justify the burgeoning excitement in this field. They also point, however, at the need for standardization of experimental design in this field and for development of new quantitative assays that must be developed to consolidate the gains that have been made.

In this chapter we provide preliminary answers to certain basic questions often posed about induced organ regeneration of adult organs. Although detailed experimental procedures have been demonstrated and a few regenerative protocols are currently being practiced clinically, there are major questions about the mechanism of the process. Just how is the spontaneous healing response modified following a regenerative treatment such as implantation of a cell-seeded scaffold? Is the process of induced regeneration in the adult essentially a recapitulation of the fetal regenerative response or is it an entirely novel healing response? What are the minimal manipulations and simplest reactants that can induce regeneration? How faithfully does a regenerated organ resemble the native organ? Most of these questions can be answered today [5]. In this chapter we summarize the salient facts about induced regeneration of three organs and explore the mechanism that explains the available data.

## 2 Nonregenerative Tissues. The Central Problem of Induced Organ Regeneration

In order to appreciate better the process of induced organ regeneration it is necessary to review the basic features of the adult healing response when it occurs spontaneously, i.e., in the absence of an inductive agent. These features have been studied in great detail at the cellular and molecular biological level where many details of the inflammatory response to injury have been documented. In this section we do not review the mechanism of spontaneous healing; instead we focus on the outcome of the injury.

Severe injury to any organ causes a discontinuity in organ mass and initiates a process of substantial loss of physiological organ function. Injury is followed swiftly by a spontaneous healing response that eventually leads to closure of the discontinuity, typically within several days. The focus here is on the outcome of the healing process, i.e., the identity of tissues that have been synthesized to close the injured site. In the mammalian fetus during the early stages of gestation, as well as in a few amphibians throughout all or part of their life, healing is a mostly reversible process that closes the discontinuity by synthesis of tissues that restore the original structure and function (spontaneous regeneration). However, in the adult mammal, healing following severe injury is a largely irreversible process that closes the injured site by organ deformation (contraction) and formation of scar, a nonphysiological tissue; this is the classical outcome of a repair process. Restoration of adult organ function following severe injury is typically only partial, occasionally threatening survival of the organism itself and necessitating organ replacement. In certain organs injury appears to be irreversible when it exceeds a critical size while, in others, it is irreversible only when it damages certain types of tissues [5]. Although the outcome of chronic injury can be described in similar terms as with acute injury, this chapter will be limited to a discussion of healing following acute injury, the area of study in which the vast majority of data have been collected so far.

Similarities in the spontaneous healing response of two organs emerge when the description is couched in terms of the individual tissues that comprise an organ. A useful classification of tissues comprising most organs includes just three types, as presented in standard textbooks of pathology [13, 14]: epithelia, basement membrane and stroma. This configuration will be referred to as the "tissue triad" [5]. Epithelia (epithelial tissue) typically cover organ surfaces in the epidermal, endocrine, genitourinary, respiratory or gastrointestinal systems, among others, and are separated from stroma by continuous basement membranes. Exceptions can be mentioned: e.g., in the liver, hepatocytes lack a basement membrane (BM); in the central nervous system, only blood vessels have a basement membrane. The composition of each tissue differs sharply from that of other members of the triad. Epithelia comprise cells and lack extracellular matrix (ECM) or blood vessels; in contrast, BMs (occasionally referred to as basal laminae) are free of cells and blood vessels; they comprise entirely of ECM. Stroma (connective tissue, supporting tissue) is the most complex of the three members of the triad, comprising, as it does, cells, blood vessels as well as ECM.

In skin, the epidermis is the epithelial tissue; it is attached to a basement membrane which, in turn, is attached to the dermis (stroma). In the peripheral nervous system (PNS), there is convincing evidence [15] that Schwann cells play the role of epithelial cells: they are capable of synthesizing a specialized, totally cellular, tissue (myelin sheath) that has a foothold on a basement membrane. The BM separates the Schwann cell-axon units (nerve fibers) from a vascularized extracellular matrix, the endoneurial stroma. Keratinocytes in the skin epidermis are "polarized", exhibiting a side that is attached to the basement membrane and another that forms the earliest tier of the maturation gradient that characterizes the epidermis; likewise, Schwann cells are also polarized, with one surface directed toward the basement membrane (abaxonal side) and the other devoted to axonal contact (axonal side) [15]. Accordingly, the epidermis-BM-dermis sequence in skin finds its counterpart in the myelin sheath-BM-endoneurial stroma sequence of the PNS [5].

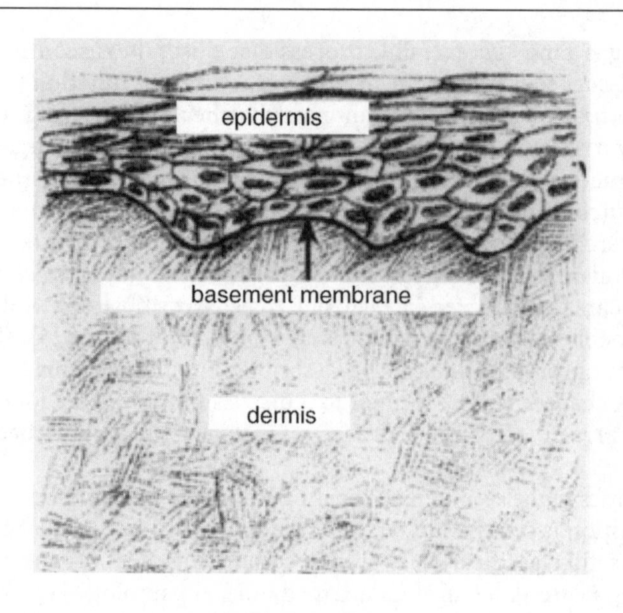


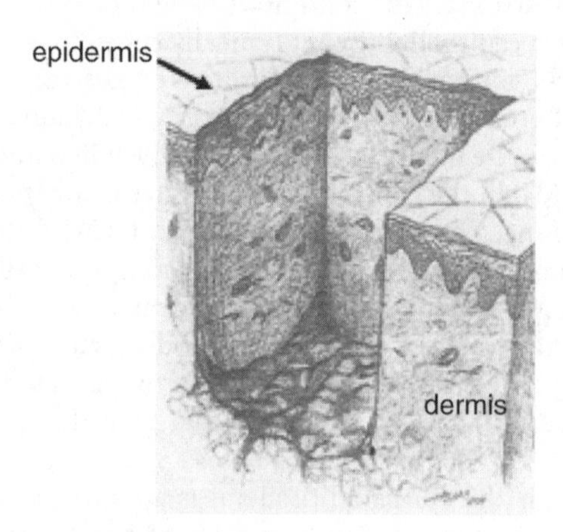


**Fig. 1** Experimental configuration for study of partial skin regeneration. *Top*: The tissue triad in skin comprises the epidermis (epithelial tissue), basement membrane and the dermis (stroma). While the first two tissues regenerate spontaneously, the dermis is non-regenerative. The central problem of induced skin regeneration is regeneration of the dermis. *Bottom*: An anatomically well-defined defect in skin, prepared by complete excision of epidermis and dermis (full-thickness skin wound), provides a suitable experimental space for study of dermis regeneration. Cells, protein solutions and scaffolds can be experimentally grafted in that well-defined space to identify reactants that are required to induce synthesis of skin (adapted with permission from [5])

There is strong evidence that, in response to injury, both the epidermis in skin and the myelin sheath in the PNS are regenerated spontaneously and, while doing so, epithelial cells also resynthesize their associated BM. These tissues are referred to as regenerative. In sharp contrast, neither the dermis nor the endoneurial stroma are regenerated spontaneously and are referred to here as nonregenerative tissues [5]. Trans-organ rules can be based on such data. The term regenerative similarity has been used to characterize tissues in different organs that heal spontaneously in a similar manner, i.e., either by regeneration or by repair. The evidence reviewed above shows that the epidermis and the myelin sheath are regeneratively similar [5]; and so are the dermis and endoneurial stroma in these organs. The regenerative nature of skin BM has been demonstrated separately [16]. It is clearly suggested from the above that, in studies of induced regeneration of an organ, the central problem is regeneration of the stroma, apparently the only nonregenerative tissue [5].

These considerations show that experimental studies of induction of organ regeneration cannot lead to clear conclusions unless the anatomical site (the experimental space) has been prepared by scrupulous elimination of all nonregenerative tissue, i.e., the tissue type to be synthesized. In addition, the experimental space should be marked by unambiguous boundaries; it should also be physically contained to prevent loss of exudates flow as well as protect from entry of neighboring tissues and infection from bacteria. Detailed rules for selecting the experimental space for such a study define the "anatomically well-defined defect" and have been presented elsewhere [5]. In what follows, the term "defect" will be consistently used to refer to an anatomically well-defined wound that is appropriate for study of induced organ regeneration (Fig. 1). In contrast, the term "wound" will be used in a much more general way to denote an injured site without further specification of the precise nature or extent of injury. Unless otherwise specified, all quantitative data presented in this chapter have been obtained with anatomically well-defined wounds, i.e., defects.

## 3 Summary of the Evidence for Induced Organ Regeneration in Adults

In this chapter we focus on three adult organs that have been induced to regenerate partially using biologically active scaffolds (regeneration templates) (e.g., see Fig. 2). As discussed in the preceding section, the protocol used to prepare the anatomical sites where these observations were made is a critical aspect of the interpretation of data. These anatomical sites where partial regeneration was induced were: 1) full-thickness skin wounds, with epidermis and dermis completely excised, in the adult guinea pig, adult swine and adult human; 2) full-thickness excision of (eye) conjunctiva, with complete excision of the stroma, in the adult rabbit; 3) the fully transected rat sciatic nerve, with stumps initially separated by a gap of unprecedented length (as of 1985!) of 15 mm; later, the gap along which regeneration of a conducting nerve was

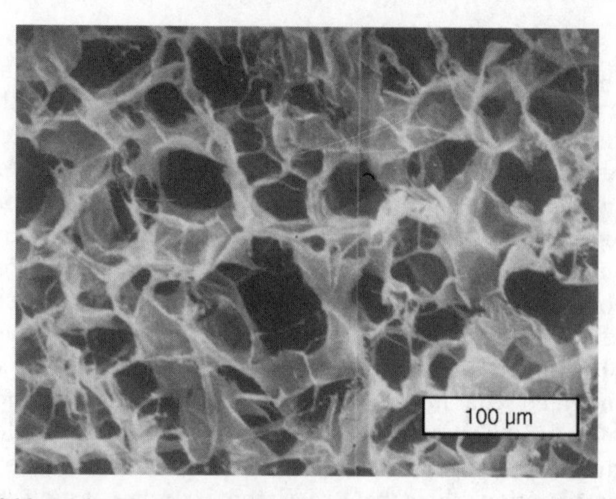


**Fig. 2** A scaffold with highly specific structure that has induced regeneration of the dermis in animals and humans (dermis regeneration template). Composition: 98/2 graft copolymer of type I collagen and chondroitin 6-sulfate; degree of residual collagen fiber crystallinity (residual banding) ~5%; average molecular weight between crosslinks, 5–15 kDa; average pore diameter, 80 μm; pore volume fraction >0.95; pore channel orientation approximately random. See Table 6. Scanning electron micrograph (courtesy of MIT)

**Table 1** List of tissues in skin, peripheral nerves and the conjunctiva that were induced to regenerate in adults

| Organ | Regenerated tissues | Tissues that did not regenerate | Not assayed |
|---|---|---|---|
| Skin (guinea pig, swine, human) [a] | Keratinized epidermis. Basement membrane. Dermis. Nerve endings. Blood vessels | Appendages (hair follicles, sweat glands, etc.) | |
| Peripheral nerve (mouse, rat, cat, monkey, human) [b] | Myelin sheath. Large- and small-diameter nerve fibers. Blood vessels. Endoneurial stroma? | | Endoneurial stroma? Perineurium |
| Conjunctiva (rabbit) [c] | Epithelium Conjunctival stroma | | Basement membrane |

[a] [6–12, 17–22].
[b] [23–28].
[c] [29].

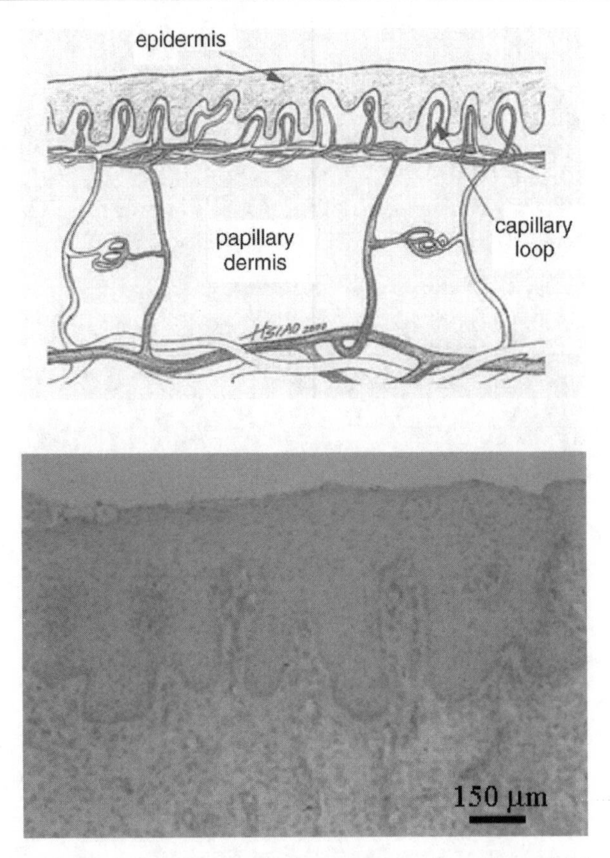


**Fig. 3** Partial evidence that regenerated skin is not scar. *Top*: Normal skin. A schematic diagram of the dermal-epidermal junction. The epidermis folds downward, forming rete ridges that interdigitate with dermal papillae, the upward folds of dermis, that enclose capillary (papillary) loops. In contrast, the interface between epidermis and dermal scar is quite flat and capillary loops are absent. *Bottom*: Regenerated skin. The dermal-epidermal junction appears very similar to that of normal skin, including the presence of capillary folds in dermal folds. Scar synthesis is ruled out. Capillaries stained with antibody to Factor VIII. Skin was partly regenerated following grafting of the dermis regeneration template (DRT), shown in Fig. 2, into a swine skin defect similar to that pictured in Fig. 1, bottom. DRT had been seeded with autologous keratinocytes (photo adapted with permission from [19])

observed was increased to about 30 mm. In all such cases it had been previously established that the excised organ does not regenerate spontaneously under these conditions, even partially, in adults; instead of closing by regeneration, these adult defects closed spontaneously by contraction and scar formation. Thus, in all these cases, an anatomically well-defined defect was used.

A summary listing of the tissues of each organ that have been induced to regenerate, as well as those that were not, is presented in Table 1.

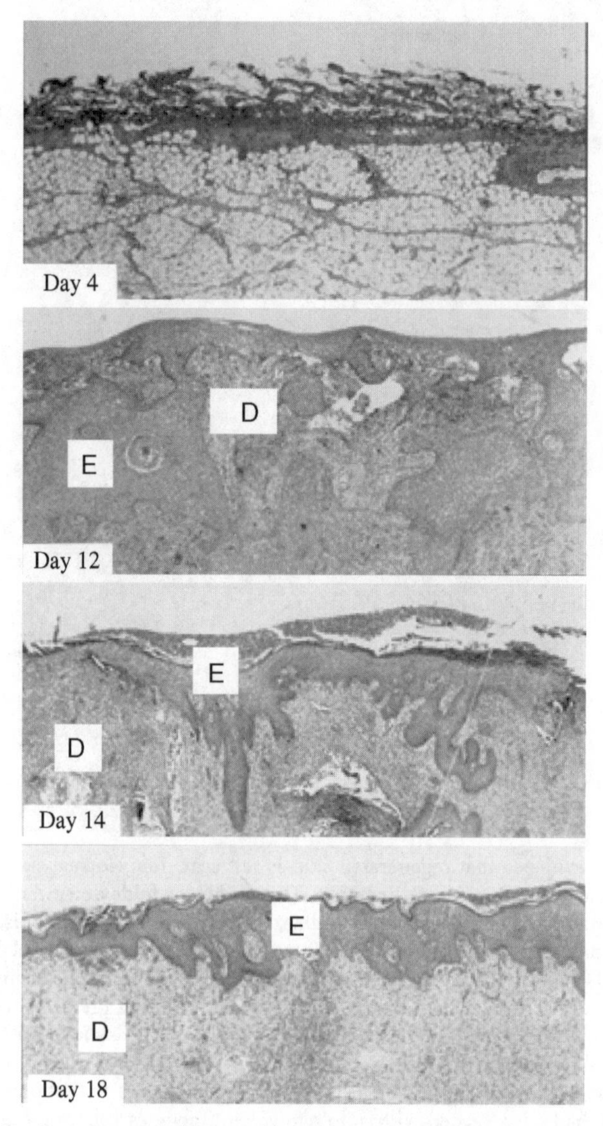


**Fig. 4** The kinetics of partial skin synthesis between 4 and 18 days. Skin regeneration was induced in the swine by grafting a full-thickness skin wound (Fig. 1) with DRT (Fig. 2) seeded with autologous keratinocytes, the only reactants that are required to achieve simultaneous synthesis of an epidermis and a dermis in the adult mammal. The cell-seeded template, shown at 4 days, is degraded almost entirely by 18 days. At 12 days, a peculiar anarchy reigns, with cords of epidermis (E) forming deep inside areas of neodermal synthesis (D). The two tissues separate out by 14 days and a near normal dermal-epidermal junction is established at 18 days (adapted with permission from [22]).

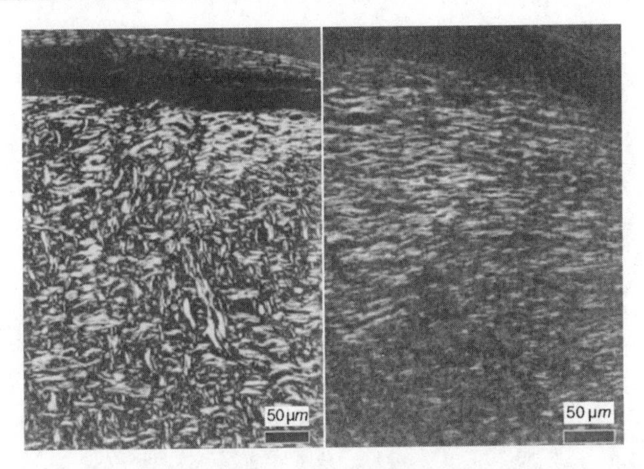


**Fig. 5** Comparison of the regenerated dermis with dermal scar. Optical microscopy using a polarizing stage was used to highlight collagen fibers, indicating their orientation. Both specimens were synthesized inside a full-thickness skin wound (Fig. 1) in the guinea pig. *Left*: Regenerated dermis. Synthesized following grafting of the keratinocyte-seeded DRT (Fig. 2). Collagen fibers are oriented almost randomly in three dimensions, similar to that in normal guinea pig dermis. *Right*: Dermal scar. Synthesized by allowing spontaneous healing of the defect (no graft). Collagen fibers have a prominent orientation in the plane of the epidermis, unlike that in normal guinea pig dermis (reproduced with permission from [5])

Briefly, confirmation of partial regeneration of *skin*, including both a dermis and an epidermis, was made by histological, immunohistochemical, ultrastructural and functional studies [6–12, 17–22]. Skin regeneration was separated from other healing processes (Orgill and Butler, in this volume). Induced regeneration was partial, i.e., perfectly physiological skin organ was not regenerated. Regenerated skin was histologically and functionally different from scar and identical to physiological skin in almost all respects, including a physiological epidermis, well-formed basement membrane, well-formed capillary loops at the rete ridges of the dermal-epidermal junction, nerve endings with confirmed tactile and heat-cold feeling, and a physiological dermis (Butler and Orgill, in this volume). However, regenerated skin lacked appendages (hair follicles, sweat glands, etc.). Some of the evidence for skin regeneration is illustrated in Figs. 3, 4, and 5.

*Peripheral nerve* regeneration was confirmed using both morphological and functional (electrophysiological and neurological) data [23–28]. Briefly, by 60 weeks, the number of axons in regenerated nerves was three times as in a normal nerve but with only half the number of large-diameter fibers (fibers with diameter exceeding 6 μm). Electrophysiological measurements showed that the regenerated nerves were able to conduct a compound nerve action potential that was similar in shape to that of a normal nerve, but with a conduction velocity that was only 75%, and an amplitude that was 36%, that of the normal nerve. Experimental conditions and some of the evidence for peripheral nerve regeneration are illustrated in Figs. 6, 7, 8, 9, and 10.

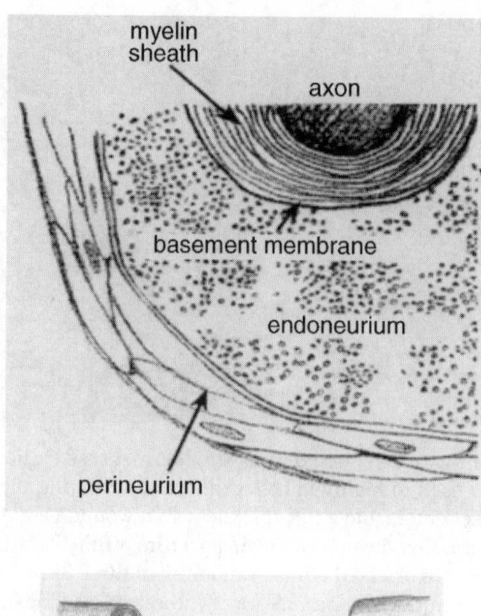

**Fig. 6** Experimental configuration for study of peripheral nerve regeneration. *Top*: In peripheral nerve, the tissue triad comprises the myelin sheath (epithelial tissue) surrounding the axoplasm, basement membrane and the endoneurium (stroma). The endoneurium is nonregenerative and is the central problem in induced regeneration of a peripheral nerve. *Bottom*: An anatomically well-defined defect in a peripheral nerve, prepared by complete transection that yields two nerve stumps (proximal and distal). The transected nerve provides a suitable experimental space for study of peripheral nerve regeneration. Most investigators have inserted the two stumps, separated by a gap of fixed length, in a tubular chamber (often a porous scaffold). Experimental efforts toward reconnection of the two stumps with partial recovery of function have centered on variations in tubular scaffold characteristics; tube fillings based on cells, soluble protein solutions and gap-filling scaffolds have also been studied (adapted with permission from [5])

Regeneration of the *conjunctiva*, including the conjunctival stroma, was observed based on histological data [29] (also Hatton and Rubin, in this volume). Regenerated stroma consisted of collagen fibers that were relatively loose and randomly oriented; when viewed with polarized light they exhibited sparse, discontinuous birefringence (characteristic of loosely arrayed crystalline collagen fibers). The organization of collagen fibers in the regenerated conjunctiva was similar to that of the stroma of the normal conjunctiva. In contrast, the stroma formed in conjunctival defects that had not been treated with the biologically active scaffold was characterized by dense packing of collagen

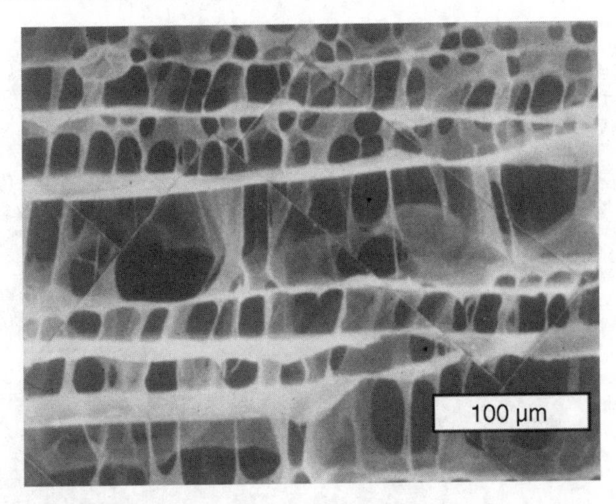


**Fig. 7** This collagen/GAG scaffold induced regeneration of the rat sciatic nerve across a 15-mm gap (and, eventually, longer gaps) when placed as a bridge between the two stumps inside a silicone tube. Regeneration did not occur in the absence of the scaffold. In recent studies the chemical composition of this scaffold has been changed to GAG-free type I collagen. Pore channel orientation along major nerve axis. Average pore diameter, 20 μm. Scanning electron micrograph. See Table 6 (courtesy of MIT)

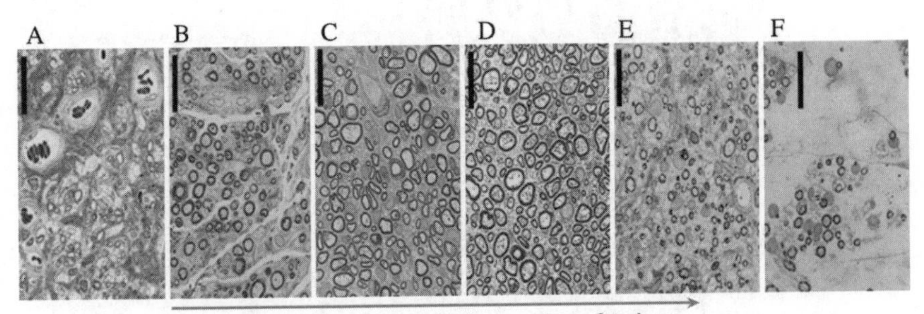


**Fig. 8** The effect of scaffold/chamber degradation rate on quality of peripheral nerve regeneration. Maximum quality of regeneration observed when the scaffold/chamber degraded with a half-life of two to three weeks. Following transection, the stumps of the transected rat sciatic nerve were inserted inside members of a series of scaffold/chambers with degradation rate that gradually decreased along the series. Cross sections of regenerated nerves were obtained at the midpoint of the regenerated nerve trunk after nine weeks. The number of axons with diameter larger than 6 μm control conduction velocity; they are a maximum in panel D, corresponding to a scaffold with intermediate degradation rate. Scale bars: ~0.25 μm (courtesy MIT)

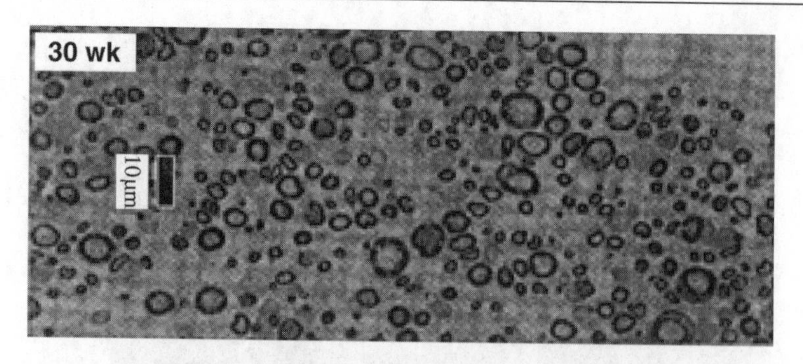
30 wk
10μm

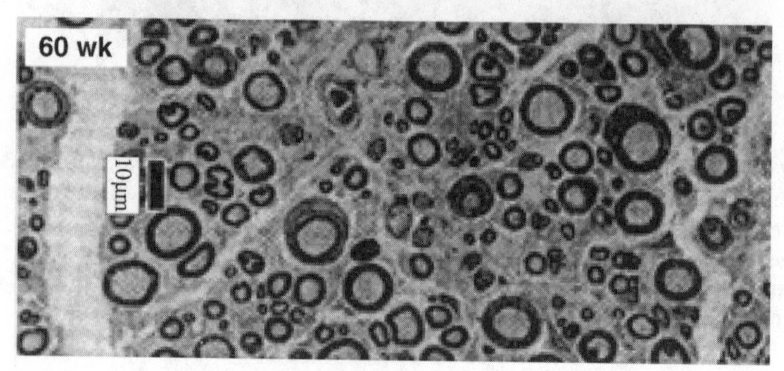
60 wk
10μm

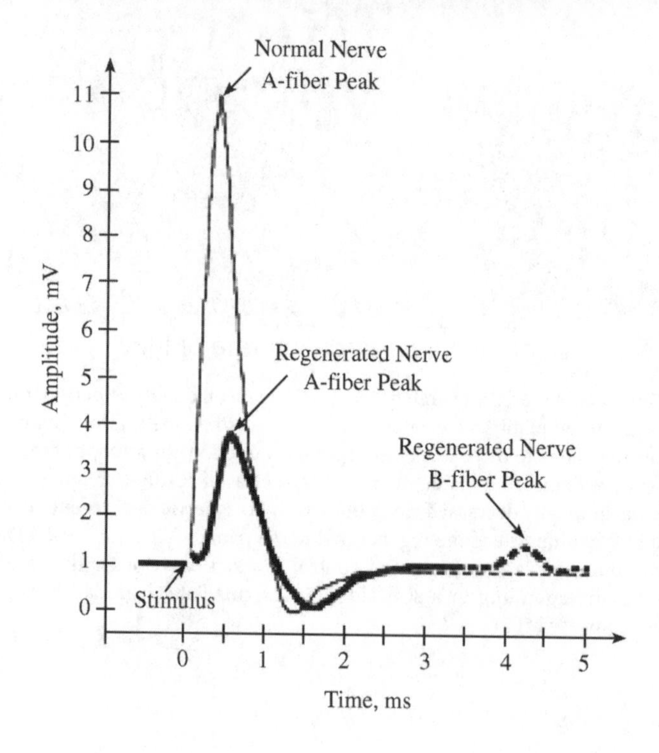
Normal Nerve
A-fiber Peak
Regenerated Nerve
A-fiber Peak
Regenerated Nerve
B-fiber Peak
Stimulus
Amplitude, mV
Time, ms
0
1
2
3
4
5
6
7
8
9
10
11

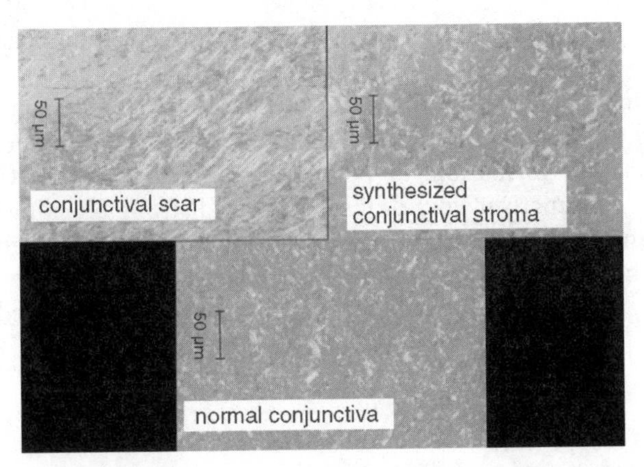


**Fig. 10** Regeneration of the conjunctiva in the rabbit. Polarized light micrographs obtained 28 days after injury (full excision of epithelial layer and stroma) illustrate differences in organization of collagen fibers in stroma. *Top left*: Conjunctival scar formed in spontaneously healing full-thickness defect. Parallel bands of birefringence show that collagen fibers were highly oriented on the epithelial plane. *Top right*: Conjunctival stroma was regenerated following grafting of the full-thickness defect with the scaffold shown in Fig. 2. Collagen fibers display birefringence with random orientation. *Bottom*: Normal conjunctival stroma shows collagen fibers that are oriented randomly, as in regenerated conjunctiva (from [29] with permission)

fibers with axes that were oriented almost entirely in the plane defined by the conjunctival epithelium and the underlying sclera; these fibers exhibited intense birefringence [29].

Observations of stroma regeneration were directly made in the case of skin and the conjunctiva; in peripheral nerve regeneration, the possible regeneration of endoneurial stroma was not directly assayed for (its potential significance had not been anticipated in these earlier studies). Indirect evidence for endoneurial stroma regeneration is available from data showing recovery of most physiological functions of peripheral nerve following use of certain partic-

**Fig. 9** *Top*: Kinetics of peripheral nerve regeneration. The rat sciatic nerve was transected and the stumps were inserted in a collagen scaffold chamber. The average diameter of axons in the regenerated nerve increased significantly from 30 weeks to 60 weeks. *Bottom*: Comparison of electrophysiological properties of a regenerated nerve with moderate quality of regeneration (*thick line*) and a normal nerve (*thin line*) at 60 weeks. Nerve regenerated inside a collagen scaffold/chamber filled with the scaffold shown in Fig. 7. The conduction velocity of the regenerated nerve was 50±2 m/s compared to 67±3 m/s for normal control. A-fiber potential corresponds to fibers larger than 6 mm in diameter; B-fiber potential observed only with regenerated nerves, corresponding to velocities about 10–25 m/s. The amplitude (equal to peak of signal) in the regenerated nerve was about 30% normal (adapted from [27] with permission).

ularly successful protocols; such functions appear to depend indispensably on the presence of an endoneurium (see review in [5]).

The available evidence in the above studies supports the conclusion that closure of these severely injured, adult anatomical sites did not occur by contraction and scar formation. In particular, studies of contraction in skin wounds and conjunctival wounds clearly showed strong blocking of contraction in the anatomical site in which regeneration was eventually demonstrated. In the peripheral nervous system, blocking of contraction of nerve regenerates was strongly hinted at by the data but was only indirectly observed [5, 30].

## 4 The Defect Closure Rule

A useful procedure for collecting and analyzing data obtained in independent studies of healing in organs is by use of the defect closure rule [5]. As noted above, the term "defect" refers to the anatomically well-defined wound. In this procedure only two time points of the defect healing process are considered: the initial and final state. Simply defined, the initial state is the recently generated defect, described by the interruption in structural continuity of one or more tissues and the start of flow of exudate, including blood, from the injured site. The final state coincides with the recent closure of the defect, an event that is followed by arrest of the flow of exudate. Even though the freshly opened and freshly closed defect in a living organism are clearly not states of static equilibrium, they are operationally defined unambiguously as two points along the experimental timescale [5].

The overall result of the healing process will now be described in terms of the total change that the defect has undergone between the initial and final states. The literature shows that no more than three processes (closure modes) are used to close a skin defect: contraction of the dermal edges of the defect, formation of epithelialized scar, and regeneration [13]. The same processes are also responsible for closure of defects in peripheral nerves [30] and in the conjunctiva [29]. The contribution of each process to closure of the defect can be measured using methodology that has been described in detail [5].

Using percentages to describe the contribution to closure of the defect area by contraction ($C$), scar formation ($S$), and regeneration ($R$), and summing up to 100, we get the defect closure rule:

$$C + S + R = 100 \quad (1)$$

This simple rule states that the final state of the healing process of a defect (closure) can be described by the contribution of just three processes to closure, namely, contraction, scar formation, and regeneration. The three numbers, enclosed in brackets by convention, are referred to as the configuration of the final state; this is a shorthand, quantitative description of the outcome of an experimental protocol. For example, the data describing spontaneous heal-

**Table 2** Representative healing processes, described according to the configuration of the final state (defect closure rule)

| Healing process | Defect closure rule symbolism |
|---|---|
| General case of organ defect healing | $[C, S, R]$ |
| Ideal fetal healing of (dermis-free) skin defect | $[0, 0, 100]$ |
| Spontaneous healing of (dermis-free) skin defect in rabbit[a] | $[96, 4, 0]$ |
| Spontaneous healing of the (dermis-free) skin defect in the adult human[b] | $[36, 63, 0]$ |
| Spontaneous healing of transected adult rat sciatic nerve[c] | $[96, 4, 0]$ |
| Spontaneous healing of stroma-free defect in adult rabbit conjunctiva[d] | $[45, 55, 0]$ |

[a] [31].
[b] [34].
[c] [5], estimated from [30].
[d] [5], estimated from [29].

ing of a full-thickness skin wound (defect) in the guinea pig can be represented as a final state of $C$=84±3, $S$=16±3, and $R$=0 (estimated from [30]); or, omitting the error signs for a simplified notation: $[C, S, R]=[84, 16, 0]$.

According to conclusions reached above, healing of a defect in the stroma of an organ in the adult mammal can be generally represented as

$$C + S = 100 \tag{2}$$

or $[C, S, 0]$ as a symbolic definition of a general repair process in the adult (absence of regeneration). Idealized healing of skin wounds in the early mammalian fetal model becomes [0, 0, 100]. Representative cases of healing are described in Table 2. In one example the data describe the well-known observations that contraction is a dominant process for closure in the rodent (due to the mobility of the integument; see [32, 33]) while scar formation plays a much more important role in the adult human. Data describing the quantitative outcome of spontaneous healing processes in various species have been reported [5].

The defect closure rule ties together quantitatively the three outcomes of healing. It will be used below to describe changes in healing processes due to experimental intervention. One of the questions that will be addressed is: When regeneration is experimentally induced in an adult, is contraction or scar formation diminished?

## 5
## The Antagonistic Role of Contraction

We review below six sets of data that describe the quantitative, or semiquantitative, relation among the three healing processes, contraction, scar formation and regeneration in various experimental models. The original data have been collected by several investigators; they have been analyzed below using the symbolism of the defect closure rule.

The first set of data derives from a study of healing of a skin defect in an amphibian (anuran) model (*Rana catesbeiana*). The model has provided direct and continuous observation of the full-thickness skin defect at different stages of development of the tadpole to an adult frog. Although opinions vary about whether amphibian data are directly applicable to mammalian healing behavior, amphibian behavior cannot be neglected in a general overview of data on healing behavior. Furthermore, it is worth noting that the tadpole offers an opportunity for collecting numerical data that has not been available with the mammalian fetus. In this study each of the three modes of closure of the dermis-free defect was measured directly at four distinct stages of tadpole development and the data are entered in Table 3. The data show that, during tadpole development, contraction and regeneration are the only two modes of closure; the tadpole data can be represented by

$$C + R = 100 \qquad (3)$$

With increasing development, contraction becomes clearly dominant at the expense of regeneration. A small component of scar formation is first observed after metamorphosis of the tadpole to the adult frog; at this adult stage, regen-

**Table 3** Modes of closure at different developmental stages. Dermis-free defects in an amphibian[a]

| Developmental stage[b] | % Contraction ($C$) | % Scar ($S$) | % Regeneration ($R$) |
|---|---|---|---|
| Larva (tadpole), premetamorphic stage | 40.8±6.8 | 0 | 59.2±6.8 |
| Larva, early prometamorphic stage | 62.1±3.0 | 0 | 37.9±3.0 |
| Larva, mid prometamorphic stage | 66.3±8.1 | 0 | 33.7±8.1 |
| Larva, late prometamorphic stage | 90.1±2.3 | 0 | 9.9±2.3 |
| Adult frog | 94±4 | 6±4 | 0 |

[a] Data for the North American bullfrog (*Rana catesbeiana*) from [35].

[b] Developmental staging based on classic staging criteria for *Rana pipiens* relating chronological age and total body length [36].

eration has been abolished and contraction accounts for almost all of closure of the defect [35], consistent with the generalized repair process in the adult mammal with a mobile integument described above.

A second set of data describes the fetal-to-adult transition in healing processes of mammalian defects. For most species studied so far, including the human, there is evidence of a developmental transition, sometime during late gestation, leading from healing primarily by regeneration to healing by repair. There is a shortage of numerical observations, occasioned by the experimental difficulty of studying the healing mammalian fetus. A description of fetal healing behavior is further complicated by observations that not all organs in a given species go through the fetal-to-adult transition at the same developmental stage [37]. A detailed discussion of fetal organ regeneration appears elsewhere (Colwell et al., in this volume). We conclude that the fetal-to-adult transition is generally characterized by a change from closure by contraction (*C*) and scar formation (*S*) to regeneration (*R*).

The third set of data that describes the relation between regeneration and contraction has been collected from a large number of studies of spontaneous healing conducted at various levels of the severity of the defect. In these studies the severity of organ injury increased progressively by use of experimental treatments that led to loss of the epithelia all the way to loss of the stroma. In skin, the ability of the epidermis to regenerate spontaneously, while the dermis does not, have both been well documented (see literature review in [5]). A similar phenomenon has been also documented with injuries in peripheral nerves. Here, following a mild crush that separates the axoplasm and the myelin sheath, but not the surrounding BM or the endoneurial stroma outside the BM, axoplasm and myelin are displaced out of the crushed site while the BM and the endoneurial stroma remain intact. Following release of the mild crushing force, the myelin sheath is regenerated completely [38, 39]. In contrast, following complete transection of the nerve, a severe injury in which the endoneurial stroma is also damaged, healing occurs by contraction [5, 30] and neural scar (neuroma) formation [30, 40]. Both in adult skin and peripheral nerves, healing of the mildly injured organ (injured epithelia but not stroma) heals by regeneration while the severely injured organ (injured stroma) heals by contraction and scar formation. We conclude that, with increase in severity of organ injury, regeneration is replaced by contraction as the dominant mode of closure.

A fourth set of data (Table 4) describes the induction of partial regeneration of skin, peripheral nerves and the conjunctiva using biologically active scaffolds (regeneration templates). A summary of the results of induced regeneration processes has been presented in Table 1. Briefly, in studies with skin and conjunctiva, direct evidence was obtained that the stroma had been regenerated using appropriate scaffolds. Indirect evidence for regeneration of the endoneurial stroma of peripheral nerves was obtained. Table 4 presents data (see review of the literature in [5]) which show that grafting of defects in skin, peripheral nerves and the conjunctiva with biologically active scaffolds leads

**Table 4** Effect of scaffold application (grafting) on healing behavior of defects in skin, peripheral nerves and the conjunctiva (see Table 1 for references; detailed review of the literature in [5])

| Defect/Species/ Organ | Scaffold used | Spontaneous healing, [*C*, *S*, *R*] | Treated with scaffold, [*C*, *S*, *R*] |
|---|---|---|---|
| Excised skin in guinea pig[a] | Dermis regeneration template (DRT) | [91, 9, 0] | [89, 0, 11] |
| Excised skin in guinea pig[b] | DRT seeded with autologous keratinocytes | [92, 8, 0] | [28, 0, 72] |
| Transected sciatic nerve in rat[c] | Silicone tube filled with nerve regeneration template (NRT) | [95, 5, 0] | [53, 0, 47] |
| Transected sciatic nerve in rat[d] | Collagen tube filled with NRT | [95, 5, 0] | [0, 0, 100] |
| Excised conjunctiva in rabbit[e] | DRT | [45, 55, 0] | [13, 0, 87] |

[a] [6–12].
[b] [6, 8–12, 18–22].
[c] [27, 28, 30].
[d] [27, 28, 30].
[e] [29].

to a sharp drop in the contribution of contraction to defect closure (*C*) and a corresponding increase in the contribution of regeneration (*R*).

The fifth set of data informs about the relation between contraction and scar formation. A distinction between the two is possible based on the numerical values of *C* and *S* reported in Table 4. The data show that a reduction in the magnitude of *C*, even when modest, is accompanied in all five cases by a drop of *S* to zero or near zero. It can be argued that a different experimental study, conducted not with rodents but with other species, with which closure by scar is comparatively more important, would show that a finite mass of scar was conceivably formed even when contraction had been blocked. Nevertheless, the rodent data generally show that scar formation, a closure mode always secondary in magnitude to contraction in spontaneously healing skin defects in rodents (Table 2), drops to near-zero values when contraction is blocked, even partly, using specific scaffolds (Table 4). The rodent data support a view of contraction as the main engine of defect closure while scar formation is viewed as a derivative (secondary) process. A mechanistic interpretation of this phenomenon appears below.

The last set of data describes the spontaneous healing behavior of defects characterized by a systemic impairment in the healing process that interferes with contraction. These data provide interesting insight into the relation

**Table 5** Summary of data relating direction of change in individual closure modes to changes in certain experimental conditions[a]

| Changes in experimental conditions | Defect | Species | Resulting changes in closure modes |
|---|---|---|---|
| 1. Tadpole development | Full-thickness skin defect | Anuran | $\Delta C>0, \Delta R<0$ |
| 2. Fetal-to-adult healing transition in mammal | Full-thickness skin defect, other organs | Various mammals | $\Delta C>0, \Delta R<0$ |
| 3. Severity of injury increases across tissues of organ | Increasing severity of injury | Various adult mammals | $\Delta C>0, \Delta R<0$ |
| 4. Apply scaffold with specific structure (regeneration template) | Fully excised skin or conjunctiva; transected nerve | Adult guinea pig, swine, human | $\Delta C<0, \Delta R>0$ |
| 5. Apply scaffold with specific structure (regeneration template) | Fully excised skin or conjunctiva; transected nerve | Adult guinea pig, swine, human | $\Delta C<0, S=0$ |
| 6. Impaired healing due to steroid treatment, genetic diabetes, genetic obesity, or infection of defects | Fully excised skin | | $\Delta C<0, R=0$ |

[a] References in the text.

between contraction, scar and regeneration. The use of steroids (e.g., cortisone acetate) led to significant inhibition of contraction in two studies [41,42]; however, no regeneration was observed. Study of healing of dermis-free defects in the diabetic mouse, a well-known model of a genetic disorder leading to impaired healing, showed that the contraction was severely suppressed relative to the normal control and that regeneration was not observed [43]. Likewise, contraction was severely suppressed in three other models of impaired healing: the genetically obese mouse [43], in contaminated defects [44] and in mechanically splinted defects [31]. Regeneration was not reported in any of these studies. The data from several studies of impaired healing show clearly that inhibition of contraction alone did not suffice to induce regeneration.

A qualitative summary of the changes in values of $S$ (scar formation) and $R$ (regeneration) observed when contraction is either up- or downregulated appears in Table 5. Based on the first four sets of data we conclude that an increase in $C$ ($\Delta C>0$ )was accompanied by a decrease in $R$ ($\Delta R<0$); conversely, a decrease in $C$ was accompanied by an increase in $R$. The data are limited and it can be argued that changes in the magnitude of contraction and regeneration processes both may result from a deeper cause. However, an alternative interpretation is the presence of an antagonistic relation between contraction and regeneration. The available data (Table 5) further suggest that this presumptively antagonistic relation appears to extend across a number of species and developmental stages. The fifth set of data shows that scar formation was abolished when contraction was suppressed even moderately. However, the sixth set of data clearly shows that blocking of contraction due to impairment in healing was not accompanied by regeneration.

By focusing on the change in contribution of each closure mode due to the changes in experimental conditions, we obtain a shorthand representation of the data in Table 5:

$$\Delta R > 0 \text{ and } S \to 0 \text{ requires } \Delta C < 0 \quad (4)$$

Blocking of contraction ($\Delta C < 0$) appears to be required for induced regeneration ($\Delta R > 0$) and the concomitant abolition of scar ($S \to 0$), but it is not sufficient since regeneration was not induced when contraction was impaired in several models of impaired healing.

## 6 Structural Determinants of Scaffold Regenerative Activity

Scaffolds are highly porous, typically macromolecular, solids that are used, either seeded with cells or unseeded, in protocols designed to synthesize tissue or organs in vitro or in vivo [45]. Two among these (referred to above as regeneration templates) have induced partial synthesis of new functional organs at the correct anatomical site and form the basis of medical devices that are currently used clinically (skin, peripheral nerves). Most biologically active scaffolds have been simple analogs of the extracellular matrix (ECM), consisting either of deliberately modified type I collagen (to reduce banding or introduce crosslinks) or graft copolymers of modified type I collagen and a glycosaminoglycan (GAG). Scaffold structure is illustrated in Figs. 2, 7, and 11.

Evidence from studies with such scaffolds has shown that the regenerative activity of templates disappears relatively rapidly when certain structural features are modified slightly, suggesting that the activity is highly specific. The sensitivity of the regenerative activity to structural modification suggests that the biological activity of templates is highly specific. Structural features of templates that account for the observed activity include the degree of selective melting of the quaternary structure of collagen, incorporation of certain ligands for fibroblasts and myofibroblasts, the presence of a minimal density of such ligands and the duration of the macromolecular network in an insoluble state over a critical period of time following implantation.

The structural features that determine regenerative activity of two scaffolds, one of which has induced regneration of skin while the other has regenerated peripheral nerves, are summarized in Table 6. Since scaffold activity appears

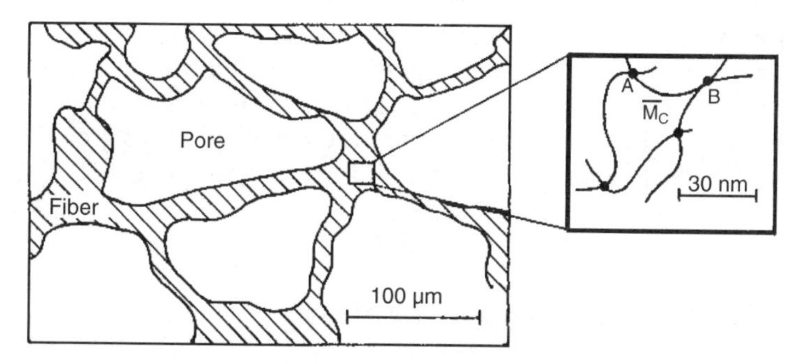


**Figure 11** Schematic illustrating the critical structural features of scaffolds. The chemical composition determines ligand identity. *Left*: The pore size (and, separately, the pore volume fraction) determines the specific surface; the latter controls ligand density. The orientation of pore channels controls the spatial coordinates of ligands. *Inset*: The crosslink density of the macromolecular network (inversely proportional to the average molecular weight between crosslinks, $M_c$) partly determines the duration of the scaffold as an insoluble matrix during the wound contraction process (courtesy of MIT)

**Table 6** Structural features of two ECM analogs with significant regenerative activity (regeneration templates) (see Fig. 11)

| Structural parameter of ECM analog with regenerative activity | ECM analog that induces | ECM analog that induces |
|---|---|---|
| | Dermis regeneration[a] | Peripheral Nerve regeneration[b] |
| Type I collagen/GAG, w/w | 98/2 | 98/2 |
| Degree of residual collagen fiber crystallinity (residual banding)[c] | ca. 5% of native collagen | ca. 5% of native collagen |
| Average molecular weight between crosslinks[d], $M_c$, kDa | 5–15 | 40–60 |
| Average pore diameter[e], μm | 20–120 | 5–10 |

[a] [9].

[b] [25, 26]. ECM analog was used as "filling" in silicone chamber. When used recently as a scaffold/chamber, the ECM analog was GAG-free.

[c] Crystallinity of collagen fibers (quaternary structure, "banding periodicity"; estimated by electron microscopy) was largely abolished following prolonged treatment in aqueous acetic acid [46]. Loss of banding of collagen fibers is accompanied by loss of their platelet aggregating activity [47].

[d] Inverse measure of crosslink density of macromolecular network. The degradation rate increases, and the diffusivity of scaffold particles increases, with an increase in $M_c$. Both an upper and a lower limit in crosslink density were required for regeneration [5, 45]. The same ECM analog was also used in regeneration of the conjunctiva [29].

[e] Both lower and higher limits of pore size were required for skin regeneration [5, 9]. See Fig. 12. Rough estimate of limits for peripheral nerve regneration [5, 25, 26) are reported here.

to be restricted to the insoluble state of a scaffold (see entry in Table 6 related to crosslink density that controls degradation rate and hence diffusivity of molecular fragments), these regulators are referred to as insoluble or nondiffusible, thereby distinguishing them from the vast majority of growth factors (e.g., TGF-β1, PDGF) which are soluble, diffusible regulators.

## 7 Mechanism of Scaffold Regenerative Activity

Just how do scaffolds induce regeneration of tissues and organs? The empirical data show that, at minimum, two steps appear to be required for induced regeneration: blocking of contraction following injury at an anatomical site (Eq. 4) and synthesis of the new tissue or organ at that site. The proposed mechanism must account for both.

Using the established structural determinants of these two templates as probes of activity, together with ancillary data, it is possible to construct a provisional mechanism for induced regeneration.

### 7.1 The Template Blocks Contraction of Injured Sites

There is strong evidence that templates block contraction very effectively in skin [7–9], peripheral nerves [30] and the conjunctiva [29]. It is not required that the template be seeded with cells. The earliest observations showed a strong delay in onset of contraction in the guinea pig full-thickness skin wound by as much as 19 days, while also unexpectedly providing a glimpse of regenerated dermis [7]. The magnitude of contraction inhibition by dermis regeneration template (DRT) in the early observations of contraction was highlighted by the use of a rodent model. Due to the mobility of their integument [32, 33], contraction is the dominant wound closure mode in rodents (Table 2) and the contraction-blocking effect of the template accordingly becomes easier to observe (Table 4).

Data from healing of peripheral nerve wounds show that contraction is equally important as in skin wounds (Fig. 12) in the closure of wounds of stumps resulting from transection (neuroma formation) [30]. Furthermore, there is substantial evidence suggesting that contraction restricts the growth of the regenerating nerve and that, by selection of the appropriate scaffold/chamber that appears to reduce greatly the density of myofibroblasts, the quality of regeneration can be greatly improved [5, 27, 30]. Data illustrating blocking of contraction by templates in skin (Fig. 13) and peripheral nerve defects (Fig. 14–16) are presented. The prominent cell type that appears in Figs. 13–16 is the differentiated myofibroblast [48] or myofibroblast (MFB), a cell type that can be identified positively by use of antibodies against $\alpha$-smooth muscle actin, that has been implicated with inducing contraction of injured sites in several

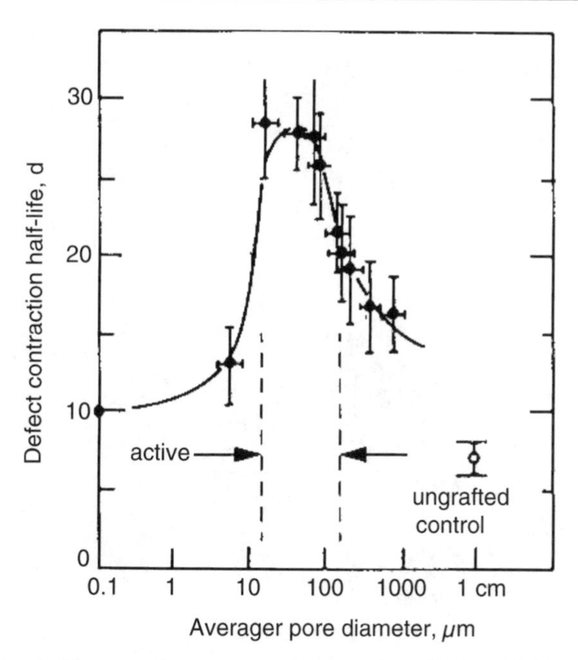


**Fig. 12** The rate of skin wound contraction for several scaffolds with identical structure except pore size. The rate reaches a minimum when the scaffold pore diameter is in the range 20–120 μm. Regenerative activity observed in this range. Structure of active scaffold that induces dermis regeneration is shown in Fig. 2 (with permission from [9])

organs in adults. Large numbers of myofibroblasts have been positively identified in injured sites, especially in template-free defects, in skin [5, 48], peripheral nerves [30] and the conjunctiva [29].

## 7.2
## Myofibroblasts, at Greatly Diminished Density, Bind on Ligands on the Template Surface

A simple way of accounting for the four structural features identified in Table 6 is by picturing the template as an unusually busy three-dimensional surface, a temporary docking station capable of binding a large number of fibroblasts and myofibroblasts over a period of time that coincides roughly with the duration of the healing process at the injured site. Each structural feature appears to play a unique role that affects the incidence of contraction blocking and the progress of tissue synthesis.

As is documented below, the diminished degree of crystallinity of the collagenous component accounts for *downregulation of the inflammatory response* at the injured site; the result is a significantly lower number of myofibroblasts. The template chemical composition determines the *identity of ligands* and selects the cell type that is bound on the surface. The specific surface of the

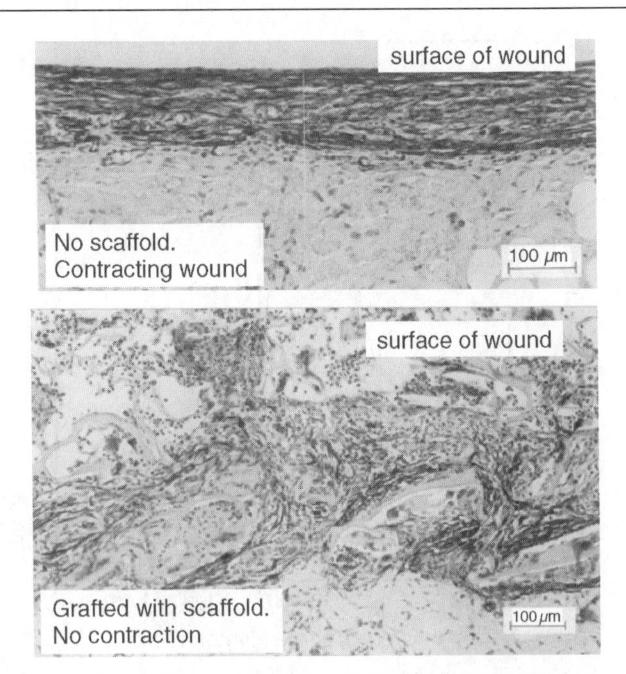


**Fig. 13** Myofibroblasts (red brown stain) observed inside a full-thickness skin wound in the guinea pig model ten days after injury. *Left*: Contraction is proceeding vigorously in this untreated skin wound. A thick, almost continuous myofibroblast layer is present at the surface of the skin wound. *Right*: Contraction has been blocked in this skin wound that has been grafted with the dermis regeneration template. Myofibroblasts are dispersed inside the template ten days after injury; cell-cell contact has been reduced and the axes of myofibroblasts are almost randomly arranged in the space of the wound. Stained with a monoclonal antibody against α-smooth muscle actin. Scale bar: ~100 μm. (courtesy of MIT)

porous network determines the *ligand density* and affects the number of cells that are bound; it is inversely proportional to the average pore diameter. The orientation of pore channels determines the *spatial coordinates of ligands* and controls the three-dimensional configuration of bound cells. Finally, the crosslink density of the macromolecular network (as well as the collagen/GAG copolymer ratio) determines the template degradation rate which, in turn, controls the *duration of the insoluble cell-matrix interface.*

Evidence that supports these detailed role assignments to the structural determinants of templates (Table 6) is available from a variety of independent sources. Supporting data are currently much more abundant for the dermis regeneration template (DRT) and the discussion below will be limited to it. Due to the relative lack of collagen banding in its structure, DRT can bind platelets but does not activate platelet degranulation [47], thereby limiting the release of inflammatory agents that facilitate differentiation of fibroblasts to myofibroblasts (MFB). Reduction in the release of factors that induce differentiation to MFB (principally, TGF-β1; see [48]) probably accounts for the observation that

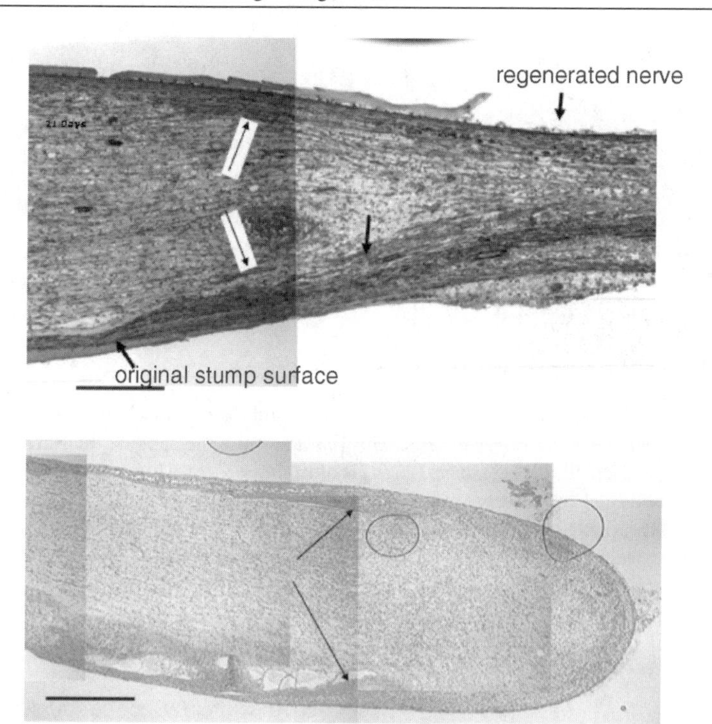


**Fig. 14** The role of contraction in healing of injured peripheral nerves. *Top*: Thick bundles of myofibroblasts (red brown stain) line the periphery of a proximal nerve stump and of the regenerated nerve 11 days after transection. The stumps were inserted inside a poorly regenerative scaffold (impermeable silicone tube); as a result, the diameter of the regenerated nerve is much smaller than normal. Longitudinal section stained with an antibody to α-smooth muscle actin (courtesy MIT). *Bottom*: Neuroma formation following transection of sciatic nerve. No scaffold was used. Longitudinal sections of the proximal stump 15 days after transection, shows contractile cell capsule (red brown stain; *arrows*) around healing stump. Stained with a monoclonal antibody against α-smooth muscle actin (courtesy of MIT)

MFB comprise only about 10% of the total number of fibroblasts in the presence of the template inside a standardized, severe skin wound, compared to about 50% in its absence [18].

Once inside the wound, fibroblasts at different levels of differentiation enter into highly organized associations with fibers at the DRT surface by extending prominent pseudopodia toward it [18]. Fibroblast-DRT binding appears to depend [51], at least in part, on participation of specific ligands, in particular those mediated by the β1-type integrins that have been shown [49] to control myofibroblast binding events during contraction. There is probably a sufficient number of sites on the DRT surface to bind most of the members of the diminished MFB population since the ligand density is high enough. Ligand density increases with the specific surface of the template; it is sufficiently large due to an average pore size that is small enough (at most 120 μm

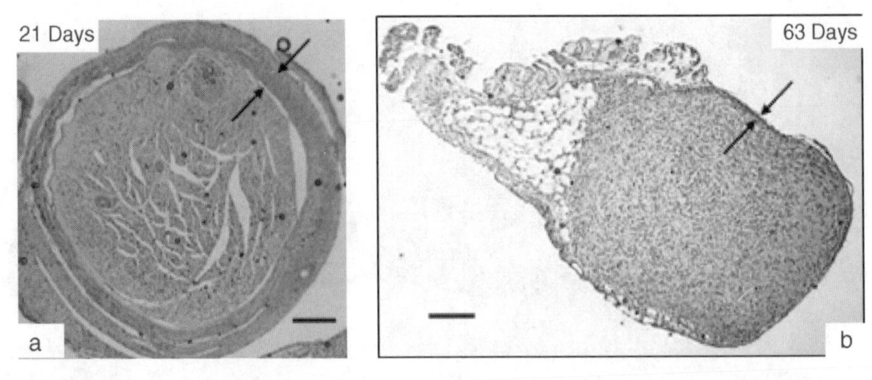


**Fig. 15** *Left*: The myofibroblast capsule (*arrows*) around the proximal stump in the silicone-tubulated nerve has a thickness of about 50–100 μm while the capsule thickness (*arrows*) around the collagen-tubulated stump (*right*) is only about 5 μm. Although the available data have been obtained at different times after implantation, the relative thickness of the myofibroblast capsule is representative of stumps tubulated by silicone tubes and collagen tubes, respectively. Stained with a monoclonal antibody against α-smooth muscle actin (red brown stain). Scale bars: ~0.25 mm (courtesy of MIT)

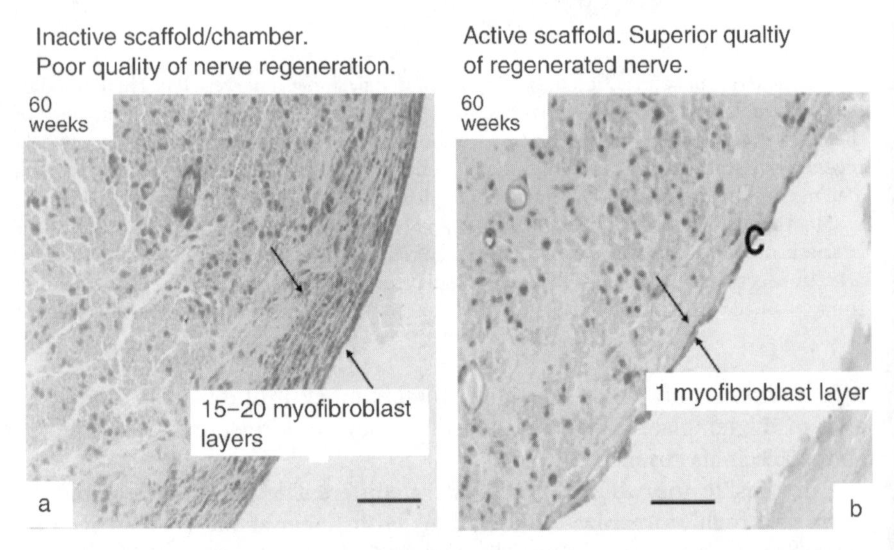


**Fig. 16** *Left*: A thick contractile cell capsule (*arrows*) surrounds a nerve trunk regenerated inside a silicone tube. *Right*: In contrast, the capsule (*arrows*; also "c") surrounding the nerve trunk regenerated inside the collagen tube is very thin. Stained with a monoclonal antibody against α-smooth muscle actin. Observed 60 weeks after implantation. Scale bars: ~50 μm (adapted with permission from [28])

for DRT; see Table 6) [52]. Although sufficiently small, the pore size of DRT is large enough (at least 20 μm; see Table 6) to ensure that MFB will be able to migrate extensively inside the template, as observed [18]. The data in Fig. 12 document the presence of a maximum in template activity in the pore size range 20–120 μm [9].

## 7.3 Effect of Pore Size on Number of Bound Myofibroblasts

The interaction of cells with the template surface can be modeled to predict that the density of bound cells must drop with an increase in average pore diameter [52].

The volume density $\rho_c$ of cells is equal to the product of the specific surface of template, $\sigma$, and the surface density of bound cells, $\Phi_c$:

$$\rho_c = \sigma \Phi_c \tag{5}$$

We will compare $\rho_c$ for two types of scaffold, differing only in average pore size.

Observations of myofibroblast density inside a template with average pore diameter of about 10 μm have yielded typical values of the volume density, $\rho_c$, of order $10^7$ myofibroblasts per $cm^3$ porous template. For this template the specific surface $\sigma$ is calculated, using a standard model for a porous solid with pore volume fraction of about 0.95, to be approximately $8\times10^4$ $mm^2/cm^3$. The cell surface density is accordingly $\Phi_c=\rho_c/\sigma=10^7/8\times10^4=125$ cells/$mm^2$. With a template of identical composition but average pore diameter as large as 300 μm, $\Phi_c$ is the same as above since the two solids have identical types of ligands and ligand density; however, the specific surface is now calculated to be about $3\times10^3$ $mm^2/cm^3$ template. In this case, the volume density of cells is only $\rho_c=\Phi_c\sigma=125\times3\times10^3=3.75\times10^5$ per $cm^3$ porous template. We conclude that the template with an average pore diameter of 10 μm binds a volume density of myofibroblasts which is about 27 times higher than does the template with a pore diameter of 300 μm. The density of ligands clearly drops as the pore diameter increases; eventually, the ligand density becomes insufficient to bind all or most of the myofibroblasts and contraction resumes, as observed [9]. These considerations suggest a maximum pore diameter requirement for the template, simply to ensure a specific surface which is large enough to bind a number of myofibroblasts that is appropriately large to block contraction across the entire scale of the defect [50, 52].

## 7.4 Minimum Duration of Template as Insoluble Matrix

The template, with MFB bound on its surface, appears to persist in a relatively undegraded state at least for about two weeks, a period approximately equal to the two- to three-week period during which contraction remains vigorous in template-free wounds [5, 7, 8]. The degradation rate is controlled by the crosslink density and by the collagen/GAG ratio (Table 6) [56, 57]. The apparent requirement for template duration that matches that of the contraction process it blocks sets upper and lower limits to the DRT degradation rate, as observed [9, 45, 58]. If the template degrades too rapidly, it fails to provide contraction blocking over the entire period during which this process is active [5]. A lower limit to the degradation rate is also observed (Table 6) and is discussed below.

## 7.5 Cancellation of Macroscopic Contraction Force

We now consider the effect of extensive myofibroblast binding on the template surface on the process of macroscopic wound contraction. In the template-free skin wound, which contracts vigorously, a very large number of myofibroblasts are organized in a very close, almost uninterrupted, association with almost all major cell axes oriented in the plane of the wound [5] (see also Fig. 13, top).

Exerting each an individual contractile force vector, $\boldsymbol{f}_i$, a number $N$ of in-plane myofibroblasts scale up the force to the level of the macroscopic force vector, $\boldsymbol{F}_c$, where wound contraction occurs:

$$\boldsymbol{F}_c = N\boldsymbol{f}_i \tag{6}$$

Measurement of the macroscopic force to contract a skin wound (in the absence of a template) yielded a force of order 0.1 N [53]. Calculation of $\boldsymbol{f}_i$ with dermal fibroblasts bound in vitro on the DRT surface led to a value of order 1 nN per cell [54, 55]. Use of Eq. (6) together with these measured values of the force per cell and the macroscopic force leads to an estimated $N=10^{-1}$ N/1 nN$=10^{8}$ cells participating in contraction of the template-free injured site. This calculation suggests, therefore, an apparent force scale-up factor of $10^{8}$ (cell→wounded organ).

In contrast, myofibroblasts bound on the template surface are much fewer (due to downregulation of the inflammatory response; see above); furthermore, their axes are mostly oriented in three-dimensional organization (out of plane of wound). Each of these features diminishes the macroscopic contractile force in the presence of the template: downregulation of the inflammatory response decreases the number of myofibroblasts in the wound and binding on the template decreases the number of in-plane MFB. Out-of-plane MFB are ineffective contributors to contraction for the following reason. There is substantial evidence that the direction of force generated by fibroblasts coincides with the direction of the long cell axis [54, 55]. Cells with axes oriented out of the plane of the wound accordingly make an individual contribution to the macroscopic force acting in the plane, $\boldsymbol{F}_c$, that diminishes with the magnitude of the angle made with the wound plane. Out-of-plane MFB therefore exert a force that is inferior to the in-plane force, $\boldsymbol{f}_i$, in Eq. (6). This configuration is consistent with near cancellation of the contribution to macroscopic contraction made by these out-of-plane cells [50]. The overall effect of the template inside the wound is, therefore, to reduce the factor $N$ and, separately, the factor

$f_i$, in Eq. (6) and thereby lead to a very significant drop in the macroscopic contraction force $F_c$ to near zero.

## 7.6
## Synthesis of Stroma

We will focus on synthesis of the nonregenerative stroma, the key event in induced skin regeneration. The focus will be on fibroblasts, the cell type that synthesize stroma [5, 59]. The synthesis of epidermis and basement membrane has been described elsewhere in detail [5].

As mentioned above, during spontaneous healing, fibroblasts and myofibroblasts are predominantly aligned in the plane of the wound [18] (see Fig. 13, top). Assuming that all or most contractile forces act in the plane of the thin layer of granulation tissue (itself in the plane of the wound), we infer that the stress distribution in the wound can be described as plane stress (out-of-plane stress components are zero). (A major contraction axis in the plane that determines the direction of movement of tissues during contraction, can nevertheless, be identified [5].)

Fibroblasts are known to deposit newly synthesized collagen fibers immediately outside their cell membrane, with the synthesized fiber axis parallel to the major cell axis [60]. It follows that, in a contracting wound, synthesis of collagen fibers takes place in the plane where the fibroblasts are oriented, yielding the fibers, highly oriented in the plane, that characterize scar (Fig. 5). Synthesis of oriented fibers proceeds in the plane provided that conditions of plane stress prevail; scar formation depends sensitively on the presence of this mechanical field. Blocking of wound contraction due to the presence of the template, marks the cancellation of the stress field that induces orientation of fibroblasts. Accordingly, in the presence of the template, the stress field is largely cancelled; fibroblasts are relaxed and are not constrained to orient themselves in the plane. Instead, they assume random orientation (Fig. 13) and synthesis now proceeds in three dimensional space, yielding a dermis with its characteristic assembly of quasi-randomly oriented collagen fibers [61] (Fig. 5).

The mechanism described here [5] explains the observed abolition of scar formation following a reduction in contractile forces in the wound (Table 5). It also explains the similarity between the topography of DRT, characterized by three-dimensional randomness in fiber orientation, and the topography of physiological stroma (dermis in skin) of the organ undergoing regeneration [59]. Such a replacement of a topography by a similar one will be referred to as being isomorphous.

## 7.7
## Synchronous Tissue Synthesis

As with the scaffold used to build a house, there seems to be a requirement for timely replacement of the template architecture by that of an ECM with similar three-dimensional structure. Does this process actually take place during regeneration?

The kinetics of dismantling of the template macromolecular structure [56] and related evidence on degradation rates of scaffolds [7, 57] support this simple model. When the ratio of the time constant for degradation (a quantity inversely related to degradation rate), $t_d$, of the scaffold was much lower than the approximate time constant for tissue synthesis during wound healing, $t_h$ (very roughly equal to 21 days for the standardized full-thickness skin wound), the scaffold did not affect significantly the contraction rate of the healing process. In contrast, when $t_d$ was much longer than about 21 days, contraction was blocked but scar formed between the intractable scaffold and the defect [7]. (This discussion also provides justification for the observed existence of a minimum in degradation rate for template activity; see Table 6.)

Synthesis of dermis (regeneration) was therefore observed when the time constants for template degradation and new tissue synthesis were approximately equal [45, 58]:

$$t_d / t_h \cong 1 \tag{7}$$

Equation (7) is a simple statement of the condition for the synchronized replacement of template by newly synthesized tissue; it can alternately be expressed in terms of the rates of the two processes involved.

When the condition of Eq. (7) is combined with the condition described in the preceding section (previously described collectively as isomorphous replacement [52]) it leads to a more general rule that will be referred to below as synchronous isomorphous placement or (SIR): the template is replaced by newly synthesized tissue (stroma) with similar architecture at about the same rate that it is being degraded.

### 7.8 Minimum Number of Reactants Required to Synthesize an Organ

In general, the cell-free template suffices to induce stroma synthesis. The synthesis of dermis in vivo can be explained without reference to cell types other than the endogenously supplied fibroblasts [5]. Nor is it necessary, as with in vitro synthetic protocols, to attempt to reproduce the cytokine field present during healing [5]; as mentioned above, the cytokine field supplied naturally by the injured site apparently provides the cell-cell signaling required for the synthesis [5]. A tremendous simplification of the protocol results from these rules of organ synthesis that apply to skin as well as peripheral nerves.

If the cell-free template alone is grafted, and the skin wound area is small enough, eventual migration of epithelial cells from the wound edge occurs, followed by synthesis of a basement membrane and a keratinizing epidermis on the newly synthesized dermal substrate (sequential organ synthesis). Although it requires use of a simpler graft, sequential synthesis of this type is lengthy; the clinical disadvantage is partly overcome by use of an autoepidermal graft to cover the newly formed dermis [17, 62]. Simultaneous synthesis of an epidermis, basement membrane and a dermis requires grafting with a template that has been previously seeded with epithelial cells (keratinocytes) [5, 56].

## 7.9
## Summary of Mechanism of Induced Skin Regeneration

The mechanism that emerges is unusual but relatively simple.

In the absence of a template the injured site closes with the spontaneous healing response. Cells of several types migrate inside the injured site while soluble proteins are being released by degranulating platelets or are being synthesized. During such template-free healing, cells and soluble signaling agents orchestrate the standard inflammatory response that leads to closure of the defect largely by contraction, dismantling of the injured matrix and eventually synthesis of a new matrix (scar) that seals the partly closed defect.

In the presence of the template the spontaneous healing response is drastically rearranged. Contraction is severely downregulated due to the observed substantial reduction in the total number of myofibroblasts (MFB) as well as extensive binding of the remaining MFB on the surface of the template (Fig. 13). These processes appear to suffice for approximate cancellation of the macroscopic mechanical force that leads to wound contraction. At the same time, however, undifferentiated fibroblasts (probably including MFB as well), bound on the random topography of the template, engage in synthesis of collagen fibers that are extruded immediately adjacent to the cell axis, generating a quasi-randomly oriented fiber array. In this manner, a new stroma is being synthesized in the contraction-free wound. Synthesis occurs at a rate that matches the rate of template degradation and in a spatial configuration that replicates the departing template. In summary, induced regeneration depends on contraction blocking (CB) followed by synchronous isomorphous template replacement (SIR).

## 7.10
## Supporting Evidence from Regeneration of Peripheral Nerve and the Conjunctiva

Although data that support this mechanism have been drawn primarily from studies of skin wound healing, there is mounting evidence that the theory accounts well for induced regeneration of two other organs as well.

In studies of the rat sciatic nerve the two stumps resulting from transection were inserted in several types of cylindrical scaffold/chambers. Treatment with an impermeable nerve chamber led to healing of individual stumps with formation of a thick capsule of MFB around the regenerated nerve (Fig. 14); the regenerate was a very thin and poorly regenerated nerve connecting the stumps. In contrast, treatment with a permeable scaffold/chamber led to formation of an extremely thin MFB capsule and a superior regenerated nerve (Figs. 15, 16). When no chamber was used each stump was surrounded by a very thick MFB and there was no reconnection of stumps (neuroma, neural scar) (Fig. 14) [30]. The data have suggested the "pressure cuff" hypothesis: During spontaneous healing of the transected nerve, the growth of the regenerating nerve is restricted mechanically by myofibroblast activity that leads to formation of hoop stresses around its perimeter; in the presence of an appropriate scaffold, myofibroblast activity is inhibited, leading to cancellation of the restrictive stress [5]. Recent studies have also shown that the quality of nerve regeneration is very sensitively dependent on the degradation rate of the scaffold/chamber in which the stumps are inserted; maximum quality of regeneration occurred when the half-life of the degrading scaffold was two to three weeks, approximately equal to the time constant, $t_h$, for healing of the injured nerve (Fig. 10) [63].

In a study of the rabbit conjunctiva it was observed that, following excision of the conjunctiva to its full depth, the untreated injury contracted vigorously and eventually closed by a combination of contraction and scar formation. In contrast, the injured site that had been treated with DRT showed greatly reduced contraction and eventually closed by formation of a regenerated conjunctiva, including conjunctival stroma [29].

## 8 Theoretical Implications of the Data

Two of the theories that have been advanced to explain the incidence of regeneration will be described briefly below.

The first theory emerges from two separate lines of inquiry. One of these is grounded on the twin observations that an unexpectedly large number of immune-related genes have been found in screens for genes expressed during regeneration; and that the immune system of many amphibians changes during metamorphosis, i.e., at the time when these amphibians lose their ability to regenerate [64] (also Mescher and Neff, in this volume).

Another line of inquiry was inspired by the relatively recent, unexpected observation that mice of the MRL strain differ from mice of other strains in one important respect: MRL mice can regenerate spontaneously the two layers of dermis (and the cartilage plate contained in between) following a perforation through the ear *pinna*. A similar observation had been made earlier with rabbits and other lagomorphs but not with the vast majority of mammals that had been studied [65, 66]. Studies of the genes expressed during healing of ear defects showed significant differences in expression of various proteases and their inhibitors in ear tissues of MRL mice and nonregenerating controls; for example, tenascin was present in MRL mice but not in controls [66]. Another study showed that the majority of genes expressed during healing of MRL mice were related to tissue growth while genes expressed in nonregenerating mice were related to the inflammatory response [67].

There is an increasing realization that the protective functions of the immune system are closely associated with mechanisms that are responsible for healing processes following injury. According to this viewpoint, failure of regeneration in limbs and other organs of anurans (frogs), reptiles and more advanced vertebrates is partly due to the highly developed status of adaptive immunity in these organisms. Vertebrates with less well-developed adaptive immunity, such as urodele amphibians that maintain the ability to regenerate throughout their entire lifetime, exhibit suppressed inflammatory responses to injury. The theory predicts that changes in immune competence are important for the gradual loss of regenerative potential that accompanies metamorphosis in anurans and that the fetal-to-adult transition in healing, usually accompanied by disabling of regeneration, is controlled by the emergence of an individual immune response [64]. Improved understanding of inability to regenerate that is observed in most organisms will, therefore, ensue from studies of cellular activities and factors released during inflammation that cause fibrosis (scar formation) and wound contraction (Mescher and Neff, in this volume).

The second theory, based on contraction-blocking followed by synchronous isomorphous replacement (CBSIR), has been described above in detail. It is based on the premise that, although suppressed during development, the fetal healing response survives in the adult and takes over when the adult response is appropriately inhibited. To be effective, the inhibition of the adult response is apparently required to be highly localized and specifically directed toward blocking of contraction; systemic methods of contraction blocking appear to be ineffective. Blocking of contraction is followed by replacement of the template by tissue (stroma) that is topographically similar and at a rate that is approximately equal to template degradation. The theory explains the available data describing the incidence of induced regeneration of three organs in a number of species of adult mammals.

Are the two theories of regeneration compatible or contradictory? The data that have led to each are quite different; as a result, each hypothesis grapples with the problem of regeneration in its own way. The first draws a parallel between the development of the individual immune system and a corresponding loss of epimorphic regenerative capability. On the other hand, CBSIR theory predicts the existence of a latent fetal healing response in the adult that may be activated following careful blocking of the adult healing response. The two hypothetical mechanisms share, in principle at least, a common basis: some of the factors involved in the immune response and in wound healing are known to be very similar, if not identical; they both predict that, if the healing response is appropriately controlled, regeneration may be induced.

The perceived similarities have an interesting consequence. The first theory explains the loss of spontaneous regenerative activity by the emergence of immunocompetence, a developmental process that replaces the fetal response. In certain instances, immunocompetence has been shown to be a reversible process since, for example, immunotolerance can be induced locally or systemically. Similarly, the second theory describes induced regeneration in the adult by the apparent reactivation of the fetal response to injury following blocking of contraction, suggesting that the fetal response has been disabled but has not been irreversibly lost, even after development to the adult.

Both theories suggest that the fetal-to-adult transition of the healing response to injury is a reversible process. If so, the fetal response must survive in a suppressed form in the adult. And, if otherwise shown to be alive and well, the hypothetically dormant fetal response can presumably be persuaded to grow back organs that have been lost.

## References

1. Hay ED (1966) Regeneration. Holt, Rinehart and Winston, New York
2. Goss RJ (1992) Regeneration versus repair. In: Cohen IK, Diegelmann RF, Lindblad WJ (eds) Wound healing. WB Saunders, Philadelphia, chap 2, p 20
3. Stocum DL (1995) Wound repair, regeneration and artificial tissues. RG Landes, Austin
4. Tsonis PA (1996) Limb regeneration. Cambridge University Press, New York
5. Yannas IV (2001) Tissue and organ regeneration in adults. Springer, Berlin Heidelberg New York
6. Yannas IV, Burke JF, Warpehoski M, Stasikelis P, Skrabut EM, Orgill DP, Giard DJ (1981) Trans Am Soc Artif Int Org 27:19
7. Yannas IV (1981) Use of artificial skin in wound management. In: Dineen P (ed) The surgical wound. Lea and Febiger, Philadelphia, chap 15, p 171
8. Yannas IV, Burke JF, Orgill DP, Skrabut, EM (1982) Science 215:174
9. Yannas IV, Lee E, Orgill DP, Skrabut EM, Murphy GF (1989) PNAS 86:933
10. Yannas IV, Burke JF, Orgill DP, Skrabut EM (1982) Trans Soc Biomat 5:24.
11. Yannas IV, Orgill DP, Skrabut EM, Burke JF (1984) Skin regeneration with a bioreplaceable polymeric template. In: Gebelein CG (ed) Polymeric materials and artificial organs. ACS, Washington, chap 13, p 191
12. Yannas IV, Lee E, Orgill DP, Ferdman A, Skrabut EM, Murphy GF (1987) Proc Am Chem Soc Div Polym Mat 57:28
13. Martinez-Hernandez A (1988) Repair, regeneration and fibrosis. In: Rubin E, Farber JL (eds) Pathology. JB Lippincott, Philadelphia
14. Burkitt HG, Young B, Heath JW (1993) Wheater's functional histology, 3rd edn. Churchill Livingstone, Edinburgh
15. Bunge RP, Bunge MB (1983) Trends Neurosci 6:499
16. Uitto J, Mauviel A, McGrath J (1996) The dermal-epidermal basement membrane zone in cutaneous wound healing. In: Clark RAF (ed) The molecular and cellular biology of wound repair. Plenum Press, New York, chap 17, p 513
17. Burke JF, Yannas IV, Quinby WC Jr, Bondoc CC, Jung WK (1981) Ann Surg 194:413
18. Murphy GF, Orgill DP, Yannas IV (1990) Lab Invest 62:305
19. Compton CC, Butler CE, Yannas IV, Warland G, Orgill DP (1998) J Invest Dermatol 110:908
20. Orgill DP, Yannas IV (1998) J Biomed Mater Res 36:531
21. Butler CE, Orgill DP, Yannas IV, Compton CC (1998) Plast Reconstr Surg 101:1572
22. Butler CE, Yannas IV, Compton CC, Correia CA, Orgill DP (1999) Br J Plast Surg 52:127
23. Yannas IV, Orgill DP, Silver J, Norregaard TV, Zervas NT, Schoene WC (1985) Trans Soc Biomater 8:146

24. Yannas IV, Orgill DP, Silver J, Norregaard T, Zervas NT, Schoene WC (1987) Regeneration of sciatic nerve across 15-mm gap by use of a polymeric template. In: Gebelein CG (ed) Advances in biomedical materials. ACS, Washington, chap 1, p 1
25. Chang AS, Yannas IV, Perutz S, Loree H, Sethi RR, Krarup C, Norregaard TV, Zervas NT, Silver J (1990) Electrophysiological study of recovery of peripheral nerves regenerated by a collagen-glycosaminoglycan copolymer matrix. In: Gebelein CG (ed) Progress in biomedical polymers. Plenum, New York, p 107
26. Chang AS, Yannas IV (1992) Peripheral nerve regeneration. In: Smith B, Adelman G (eds) Neuroscience year. Birkhauser, Boston
27. Chamberlain LJ, Yannas IV, Hsu H-P, Strichartz G, Spector M (1998) Exp Neurol 154:315
28. Chamberlain LJ, Yannas IV, Hsu H-P, Strichartz GR, Spector M (2000) J Neurosci Res 60:666
29. Hsu W-C, Spilker MH, Yannas IV, Rubin PAD (2000) Invest Ophthalmol Vis Sci 41:2404
30. Chamberlain LJ, Yannas IV, Hsu H-P, Spector M (2000) J Comp Neurol 417:415
31. Kennedy DF, Cliff WJ (1979) Pathol 11:207
32. Billingham RE, Medawar PB (1951) J Exp Biol 28:385
33. Billingham RE, Medawar PB (1955) J Anat 89:114
34. Ramirez AT, Soroff HS, Schwartz MS, Mooty J, Pearson E, Raben MS (1969) Surg Gynecol Obst 128:283
35. Yannas IV, Colt J, Wai YC (1996) Wound Rep Reg 4:31
36. Taylor AC, Kollros JJ (1946) Anat Rec 94:7
37. Longaker MT, Adzick NS (1991) Plast Reconstr Surg 87:788
38. Goodrum JF, Weaver JE, Goines ND, Bouldin TW (1995) J Neurochem 65:1752
39. Goodrum JF, Fowler KA, Hostetter JD (2000) J Neurosci Res 15:581
40. Wall PD, Gutnick M (1974) Exp Neurol 43:580
41. Billingham RE, Russell PS (1956) Ann Surg 144:961
42. McGrath MH (1982) Plast Reconstr Surg 69:71
43. Klingbeil CK, Cesar LB, Fiddes C (1991) Basic fibroblast growth factor accelerated tissue repair in models of impaired wound healing. In: Barbul A, Caldwell M, Eaglstein W, Hunt T, Marshall D, Pines E, Skover GG (eds) Clinical and experimental approaches to dermal and epidermal repair: normal and chronic wounds. Alan R. Liss, New York
44. Hayward P, Hokanson J, Heggers J, Fiddes J, Klingbeil C (1992) Am J Surg 163:288
45. Yannas IV, Burke JF (1980) J Biomed Mater Res 14:65
46. Yannas IV (1990) Angew Chem (Engl Ed) 29:20
47. Sylvester MF, Yannas IV, Salzman EW, Forbes MJ (1989) Thromb Res 55:135
48. Tomasek JJ, Gabbiani G, Hinz B, Chaponnier C, Brown RA (2002) Nature Revs Molec Cell Biol 3:349
49. Racine-Samson L, Rockey DC, Bissel DM (1997) J Biol Chem 272:30911
50. Yannas IV (1998) Wound Rep Reg 6:518
51. Sethi KK, Yannas IV, Mudera V, Eastwood M, McFarland C, Brown RA (2002) Wound Rep Reg 10:397
52. Yannas IV (1997) Models of organ regeneration processes induced by templates. In: Prokop A, Hunkeler D, Cherrington AD (eds) Bioartificial organs. Ann NY Acad Sci 831: 280
53. Higton DIR, James DW (1964) Br J Plast Surg 51:462
54. Freyman TM, Yannas IV, Pek Y-S, Yokoo R, Gibson LJ (2001) Exp Cell Res 269:140
55. Freyman TM, Yannas IV, Yokoo R, Gibson LJ (2001) Biomaterials 22:2883
56. Yannas IV, Burke JF, Huang C, Gordon PL (1975) Polym Prepr Am Chem Soc 16:209
57. Yannas IV, Burke JF, Huang C, Gordon PL (1975) J Biomed Mater Res 9:623

58. Yannas IV, Burke JF, Umbreit M, Stasikelis P (1979) Fed Proc Am Soc Exp Biol 38:988
59. Yannas IV (2004) ChemBioChem 5:26
60. Birk DE, Trelstad RL (1985) Fibroblasts compartmentalize the extracellular space to regulate and facilitate collagen fibril, bundle, and macro-aggregate formation. In: Reddi AH (ed) Extracellular matrix: structure and function. Alan R. Liss, New York, chap 22, p 513
61. Ferdman AG, Yannas IV (1993) J Invest Dermat 100:710
62. Heimbach D, Luterman A, Burke J, Cram A, Herndon D, Hunt J, Jordan M, McManus W, Solem L, Warden G, Zawacki B (1988) Ann Surg 208:313
63. Harley BA, Spilker MH, Wu JW, Asano K, Hsu H-P, Spector M, Yannas IV (2004) Cells Tissues Organs 176:153
64. Harty M, Neff AW, King MW, Mescher AL (2003) Dev Dyn 226:268
65. Clark LD, Clark RK, Heber-Katz E (1998) Clin Immunol Immunopathol 88:35
66. Heber-Katz E (1999) Semin Cell Dev Biol 10:415
67. Li X, Mohan S, Gu W, Baylink BJ (2001) Mamm Genome 12:52

Received: March 2004

Adv Biochem Engin/Biotechnol (2005) 93: 39–66
DOI 10.1007/b99966


# Regenerative Capacity and the Developing Immune System

Anthony L. Mescher (✉) · Anton W. Neff

Center for Regenerative Biology and Medicine, Indiana University School of Medicine, Bloomington, IN 47405-4401, USA
*mescher@indiana.edu*
*neff@indiana.edu*



**Abstract** Many components of the vertebrate immune system have evolved with dual, interrelated functions of both protecting injured tissues from infection and providing for tissue maintenance and repair of injuries. The capacity for organ regeneration, prominent among invertebrates and certain phylogenically primitive vertebrates, is poorly developed in mammals. We have proposed that evolution of the mammalian immune system has produced inflammatory cellular interactions at sites of injury which have optimized tissue defense and facilitated tissue repair, but that these improvements included concomitant loss of regenerative capacity.

This chapter briefly reviews work in two regenerating systems: scar-free repair of fetal mammalian skin and regeneration of amputated limbs in larval frogs. In both organs the potential to regenerate anatomically and functionally complete new structures is lost gradually during ontogeny and this loss coincides with development of an immune system producing an inflammatory response in injured tissues. Failure of organ regeneration has long been associated with scarring or fibrosis and this phenomenon is a direct result of inflammatory interactions of immune cells and fibroblasts at sites of injury.

Several aspects of immunity related to repair are reviewed, including the importance of antigen-presenting cells and lymphocytes, relevant cytokines and growth factors released by these and other cells, immune functions of extracellular matrix components, and immunological functions of fibroblasts. Skin repair in various transgenic mouse models has been especially informative. Further study of immune mechanisms associated with the loss of regenerative capacity in the skin and amphibian limb will be useful for efforts to promote mammalian organ regeneration.

**Keywords** Limb regeneration · Fetal wound healing · Scarring · Cytokine · Interleukin

**List of Abbreviations**

| | |
|---|---|
| APC | Antigen-presenting cell |
| C3 | Complement system factor 3 |
| CD40L | CD40 ligand |
| COX | Cyclooxygenase |
| CTGF | Connective tissue growth factor |
| DC | Dendritic cell |
| DETC | Dendritic epidermal T cell (aka γδ-T cell) |
| ECM | Extracellular matrix |
| FGF | Fibroblast growth factor |
| HA | Hyaluronic acid |
| IFN | Interferon |
| IGF | Insulin-like growth factor |
| IL | Interleukin |
| LC | Langerhans cell |
| MHC | Major histocompatibility complex |
| NF-κB | Nuclear factor-kappa B |
| NK cells | Natural killer cells |
| PGE | Prostaglandin E |
| TCR | T cell receptor |
| TGF | Transforming growth factor |
| TLR | *Toll*-like receptor |

"It seems that regeneration is not necessarily adaptive, but that it has disappeared or persisted as other more important physiological attributes evolved with which it may or may not have been incompatible. Hence, before any intelligent attempt can be made to restore regeneration where it does not normally take place, it will first be necessary to learn what physiological advantages took precedence over regeneration in the course of natural selection."

Richard Goss, *Principles of Regeneration* [1]

# 1
# Introduction

As the topics reviewed in this volume make clear, the capacity of organs or tissues to restore normal structure and function after removal or injury is rare in mammals. What occurs instead is a process of epithelial closure and fibrotic scarring which heals the wound, but the affected tissues never regain their original structure or complete range of functions. Even the relatively simple repair of an incisional skin wound produces a scar that lacks differentiated epidermal structures such as hair follicles and sweat glands [2].

The potential for regeneration is very different with more primitive animals. The remarkable ability of many invertebrates to regenerate complex structures or entire organisms from small pieces of themselves is well-known. Among vertebrates a well-developed capacity to regenerate anatomically complete and fully functional tissues and organs including appendages is limited primarily to fish and amphibians [3]. Although regeneration of limbs has received the most attention from developmental biologists (reviewed in [4–6]), many urodele amphibians (the newts and salamanders) can regenerate a diverse set of organs and tissues with little evidence of scarring. Well-studied examples in newts include regeneration of the tail, jaws, spinal cord, lens, myocardium, and at least some regions of the gastrointestinal tract (reviewed in [7]). Evolution of regenerative potential has been discussed by several leaders in the field, but a long-standing question has been why a property as apparently advantageous as the capacity to regenerate was not retained during the course of vertebrate evolution [1, 8–10].

Goss [1] points out that before this question can be answered and certainly before the possibility of inducing regeneration in mammalian organs can be addressed we must understand what physiological advantages took precedence over regeneration in the course of vertebrate evolution. We have proposed that evolution of the adaptive immune system led to the appearance of new cell types and new uses for signaling cascades which together changed the response of appendages and other organs to injury [11]. A key feature of this hypothesis is that cellular interactions within the newly evolved immune response in primitive vertebrates allowed greater defense against pathogens in injured tissue, but at the cost of the tissue's ability to regenerate perfectly and restore function completely.

Protection against microbial invasion is clearly an important part of wound healing in all species, but it is also now known that in injured, inflamed tissues many immune system components have equally important roles in cell survival and reparative growth. Indeed speculation on how the more specialized and adaptive immune system arose in primitive vertebrates some 400 million years ago centers on selection for an improved protective mucosal response to more frequent localized injuries and infections that likely accompanied the predatory way of life made possible by the development of hinged jaws [12]. Mechanisms for tissue repair and strategies for combating infection seem to have evolved in vertebrates as integral parts of the same system of cells and signaling mechanisms.

In this chapter we briefly review the various components of the vertebrate immune system, emphasizing their roles in the response of tissues to injury. We then discuss work in two model systems – scarless wound healing in fetal mammals and limb regeneration in anuran amphibians – where declines in regenerative capacity are seen during ontogenic development. Finally we discuss the possibility that evolutionary refinements in the systems controlling both tissue repair and immunological protection led to suppressed regenerative capacity in mammals.

## 2 Components of the Immune System have Direct Roles in the Response to Injury

While the regenerative capacity of vertebrate organs seems to have generally declined during evolution, the immune defense mechanisms against microorganisms became more diverse and highly developed. Primitive vertebrates retained and refined the antimicrobial systems of invertebrates into what is now referred to as "innate immunity." However early vertebrates also evolved new genetic mechanisms which allowed the gradual development of "adaptive immunity" based on the genetic recombination properties of lymphocytes. The new system provided greater specificity in defense against pathogens and included a memory component that allowed very rapid responses upon reexposure to a microbial invader.

### 2.1 Aspects of Innate Immunity Affecting Growth

Even the most primitive defense systems of multicellular organisms have been shown to be intimately related to tissue repair or regeneration. Surface epithelial cells from plants to humans produce antimicrobial polypeptides in response to contact with microorganisms. In insects such polypeptides are also secreted into the circulation by fat bodies, the functional equivalents of the liver. Hundreds of these antimicrobial factors in several families have been identified, most notably the defensins which act against a range of fungi, bacteria and certain viruses by permeabilizing their cell membranes [13]. Important for the present discussion is the recent discovery that besides their antimicrobial properties, these proteins also stimulate activity of phagocytic cells and exert both mitogenic and angiogenic effects that directly promote tissue repair [14]. Conversely, recent work indicates that certain mitogenic growth factors locally released in response to injury, including transforming growth factor-α (TGF-α) and insulin-like growth factor I (IGF-I), upregulate expression of defensins and other antimicrobial peptides in skin [15].

Another component of innate immunity, the complement system, provides further defense against microbes invading vertebrate tissues either by lysing them directly or marking them for phagocytosis. In mammals this system consists of some 30 circulating proteins made mainly in the liver which remain inactive until an activation cascade is triggered locally by the presence of a pathogen. Certain complement proteins and their cleavage products also act as chemokines, recruiting inflammatory cells to sites of injury. As described in a later section, recent reports show that some proteins of the complement system are also synthesized locally in various regenerating systems as well as during endochondral bone formation and certain other developmental processes. This new work suggests that complement may play previously unsuspected roles in cell survival, dedifferentiation and proliferation [16].

Invading microorganisms faced with these defensive measures are further assaulted and eliminated by specialized granulocytes arriving on the scene from the blood. The most important of these cells for bacterial removal are the neutrophils, which are usually the first immune cells to accumulate at a wound site after responding to chemoattractants from injured cells, microorganisms, and possibly resident mast cells [17]. Besides phagocytosing and killing bacteria, neutrophils also release proteases, chemokines, and growth factors which break down extracellular matrix (ECM) components, augment the inflammatory response, and stimulate further cell proliferation and migration within the injured tissues [18]. Neutrophil-derived factors no doubt help drive the repair process, but neutrophils appear to be dispensable among the multiple sources of similar factors at a wound site. Simpson and Ross [19] showed that depleting neutrophils from experimental animals with specific antiserum had no significant effect on tissue repair as long as the wound remained sterile.

Natural killer (NK) cells, still another arm of the innate immune system, are circulating blood cells that also accumulate at wound sites where they interact with invading fungal and parasitic cells, as well as virus-infected cells, to induce an apoptotic response [20]. All vertebrate cells express major histocompatibility complex (MHC) class I transmembrane surface proteins, a signal which suppresses NK cell function. Cells lacking MHC class I or expressing insufficient amounts of this protein, a common event in virally infected or transformed cells, become targets for NK cells. Like neutrophils NK cells secrete a variety of matrix metalloproteinases, chemokines, and cytokine growth factors which lead to an augmented immune response and improved tissue growth [21].

Macrophages represent an immune cell lineage with multiple roles in inflammation and tissue repair. Elicited from the blood to accumulate in wounds by chemokines released earlier during inflammation, macrophages are derived from circulating monocytes and activated by factors from both bacteria and injured cells. Macrophages engulf and remove tissue debris and dead cells, including played-out neutrophils, preparing the microenvironment for assembly of new ECM. At the same time these cells also secrete a plethora of mitogenic and angiogenic factors promoting tissue regrowth. Experimentally depleting this cell population with hydrocortisone and antiserum against macrophages

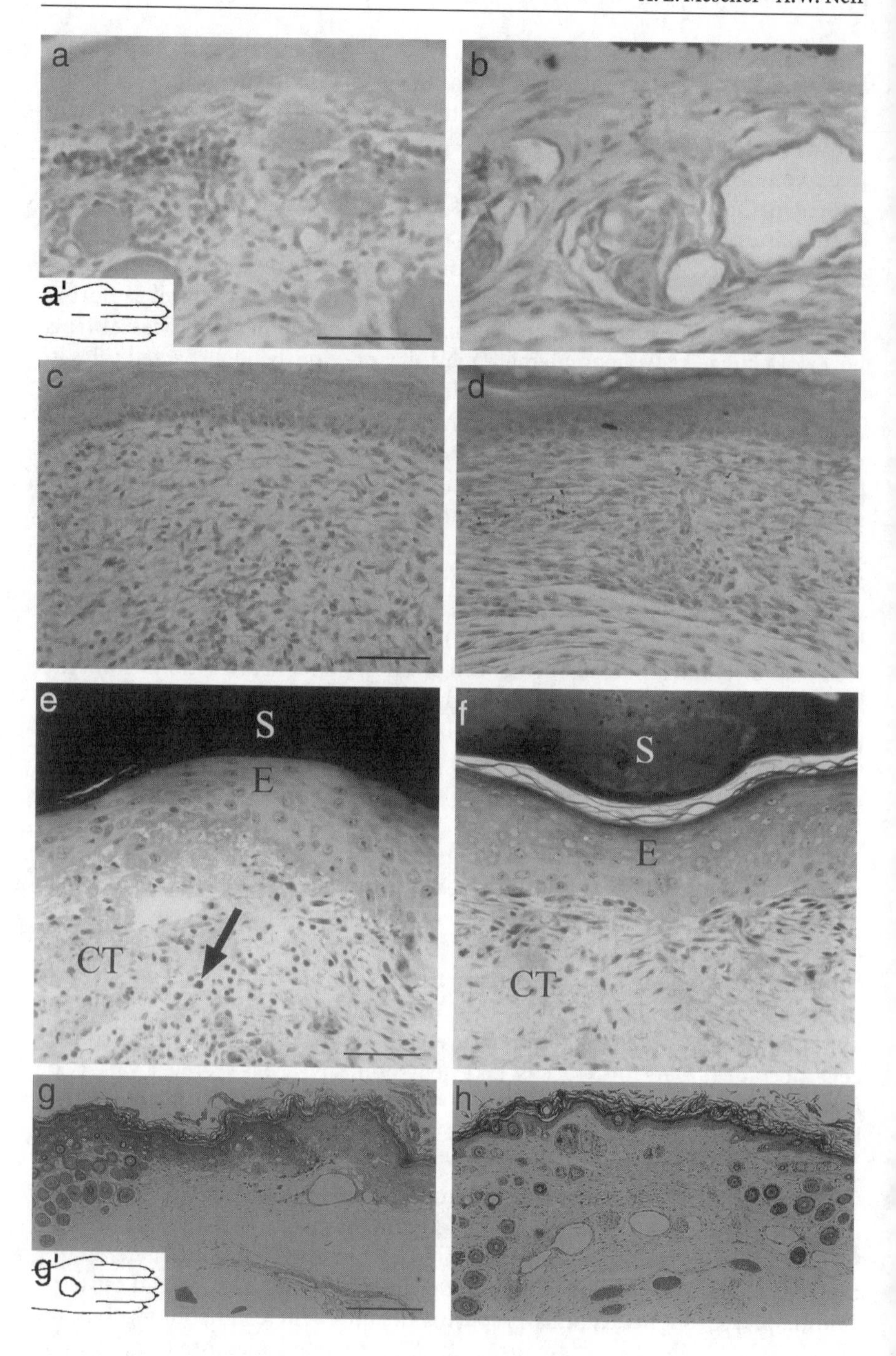
a
a'
b
c
d
e
S
E
CT
f
S
E
CT
g
g'
h

was found to retard skin wound healing [22], indicating an important role for this source of growth factors. Recently however this conclusion has been called into question by work with PU.1 null mice in which deletion of a transcription factor required to produce specific hematopoietic lineages yields mice lacking both macrophages and neutrophils [23]. Skin wounds in such mice were not only repaired at the same rate as wounds in wild-type siblings, this was achieved without inflammation or scar formation (Fig. 1). These new findings suggest that macrophages are in fact not necessary for efficient tissue repair and that the result with antiserum and hydrocortisone was caused by other effects of the steroid.

## 2.2 Adaptive Immunity and Cytokines

A key advance in the system of adaptive immune defense is the antigen-presenting cell (APC), a class of cells to which macrophages belong. In response to interferon-γ APCs are activated at sites of injury and express MHC class II proteins. MHC class II proteins on the cell surface present fragments of phagocytosed and proteosomally processed macromolecules to neighboring lymphocytes, thus initiating interactions with these cells that lead to the activation of an adaptive immune response against those specific antigens. Other APCs are more highly specialized than macrophages for antigen presentation, including

**Fig. 1a–h** Scar-free repair and reduced inflammation in PU.1 null neonatal mice lacking macrophages (from [23]); **a** numerous clustered round cells with chloroacetate esterase (neutrophil-specific) staining have been recruited to the wild-type wound tissue at one day post-wounding; **b** an equivalent one-day PU.1 null wound shows no such neutrophil recruitment; **c** the wound granulation tissue is heavily populated by F4/80-positive (shown here as *dark gray staining*) macrophages at four days post-wounding in the wild-type; **d** the equivalent wound in a PU.1 null animal is devoid of macrophages, as seen by F4/80 immunostaining; **e** resin histology of a three-day wild-type incisional wound showing the recently covered keratinocyte epithelial layer (E) beneath a scab (S). Note the epithelial layer appears immature in that there are no squames between it and the scab. In the wound connective tissue (CT), dark staining inflammatory cells and apoptotic cells (indistinguishable in resin histology) are visible; **f** an equivalent PU.1 null wound reveals a somewhat more mature, repaired keratinocyte epidermal layer (note keratin squames beneath scab), covering a relatively well-organized wound connective tissue with no apparent inflammatory cells or apoptotic corpses; **g** at seven days post-wounding, a wild-type excisional wound is fully repaired and has lost its scab, but the connective tissue at the wound site remains denser than in adjacent unwounded tissue; **h** in the equivalent PU.1 null wound, the connective tissue at the wound site appears very similar to that in adjacent unwounded tissue, with no apparent scar. a′ and g′ are schematics to illustrate the approximate size and location of standard incisional and excisional wounds made. *Scale bars* equal 20 μm for a, b, e, and f; 50 μm for c and d; and 100 μm for g and h. (Reprinted from [23] with author's and publisher's permission)

Langerhans cells of epidermis and dendritic cells (DCs) of dermis, other connective tissues and lymphoid organs. Langerhans cells and most DCs have the same myeloid origin as monocytes/macrophages, but other DCs also capable of activating T cells appear to originate in lymphoid organs [24]. All APCs are parts of the innate immune system, but with their key role in stimulating and directing lymphocytic function, they represent a major development in the evolution of the adaptive immune system in primitive vertebrates.

Activated mammalian DCs presenting infection- or injury-related markers, migrate to adjacent lymph nodes where they sensitize lymphocytes expressing specific T cell receptors (TCRs) that recognize the presented antigens. Such lymphocytes then proliferate clonally, move back to the site of inflammation, and carry out a greatly amplified immune response. Some activated macrophages stay in the wound environment where, besides their phagocytic and growth-promoting functions, they serve to restimulate and maintain the activation state of lymphocytes responding to microorganisms.

Perhaps more importantly for situations such as tissue and organ regeneration in which new antigens may be generated, DCs and other APCs also mediate a level of protection against autoimmunity termed "peripheral tolerance." APCs in the skin or other organs can induce lymphocytic tolerance for self-antigens that were not encountered by developing lymphocytes in the thymus [25]. Whether lymphocytes respond to presented antigens with non-inflammatory tolerance, aggressive cytotoxic activity, or some intermediate effect depends in part on the developmental "maturity" of DCs and the density of MHC proteins and types of co-stimulatory molecules on these and other APCs [25]. Tolerogenic interactions of APCs have been suggested as having particular importance for the outcome of events following CNS tissue damage [26].

Cells of the innate immune system have additional mechanisms for integrating the immune and tissue-repair responses, which include a variety of surface *toll*-like receptors (TLRs) that recognize all classes of invading microorganisms as well as "heat shock proteins" released from injured or necrotic (but not apoptotic) cells. Binding of such ligands to TLRs triggers synthesis of cytokines which affect other inflammatory cells, greatly augmenting their responses against the invaders and promoting tissue repair. *Toll* was first identified as a component in the signaling pathway controlling dorsal-ventral patterning in the early *Drosophila* embryo and it is not surprising that vertebrate TLR signaling has effects beyond those strictly related to immune recognition. TLR pathways of immune cells have been shown to induce expression of many developmentally important genes directly involved in tissue repair, including matrix metalloproteinases, cytokines, and angiogenic factors [27]. It is clear that innate immune cells, including the APCs which activate the adaptive immune response, have important non-immune functions in producing factors that regulate ECM remodeling and new tissue formation.

The role of lymphocytes in inflammation and tissue repair is complex and still poorly understood. Although the profile of circulating antibodies is altered by severe injuries, there is no evidence that B lymphocytes participate in tissue repair [28]. However various subsets of T cells have activities that at least indirectly influence the outcome of repair. During inflammation or infection T-helper cells are activated by DCs or other APCs and begin to produce sets of cytokines to stimulate appropriate cells at the affected site. The so-called Th1 cytokines, including interferon-γ (IFN-γ), interleukin-2 (IL-2), IL-12, IL-18, and tumor necrosis factor, produce cell-mediated immunity by acting primarily on cytotoxic T cells, macrophages and NK cells. Th2 cytokines, notably IL-4, IL-5, IL-10, and IL-13 stimulate antibody-based or humoral immunity of B cells. Because of feedback mechanisms involving specific Th1 or Th2 cytokines, synthesis of the cytokine profiles by T-helper cells tends to become mutually exclusive. This typically leads to the development of either cell-mediated or humoral immunity depending on the nature of the injury and the kinds of infectious agents initially detected. New work, further discussed below, shows that these cytokines regulate the activities of not only the local immune cells but also the activity of local fibroblasts and endothelial cells as well. Thus the cytokine profile affects not only the outcome of the infection or inflammation, but also the nature of the reparative response within the injured tissue [29, 30].

## 2.3 γδ-T Cells (Dendritic Epidermal T Cells)

Recent work suggests that another subset of mammalian T lymphocytes, those bearing γδ-TCRs, may also be very important for tissue regeneration. Unlike the well-studied T cells mentioned above, which express αβ-TCRs, γδ-T cells do not accumulate in lymphoid tissue and are found primarily in skin and the mucosae of the digestive, respiratory, and reproductive tracts, where they commonly occur as intraepithelial lymphocytes [31]. Because γδ-T lymphocytes of epidermis have characteristic dendritic morphology, they are commonly called dendritic epidermal T cells (DETCs) and it is important to distinguish them from the DCs and LCs mentioned earlier. Both of the latter are APCs, expressing MHC class II but lacking the Thy-1 surface antigen characteristic of lymphocytes. DETCs, on the other hand, express Thy-1 but not MHC II and therefore cannot process and present antigens.

Consistent with the relative absence of DETCs in lymph nodes and the spleen, these specialized T lymphocytes seem not to depend on APCs for activation against antigens. Instead it is likely that γδ-TCRs recognize "stress antigens," often low-molecular-mass nonpeptidic products, directly in affected tissues [31, 32]. Importantly, DETCs have been implicated as direct mediators of tissue repair. In wounded mouse skin these cells are quickly activated, becoming less dendritic and producing chemokines as well as keratinocyte growth factors 1 and 2 (also known as fibroblast growth factors 7 and 10) [33]. Members of the FGF family have been implicated in almost every known example of tissue or organ regeneration. TCRγ$^{-/-}$ mice lack γδ-T cells and have significantly impaired ability to repair incisional wounds, with greatly reduced keratinocyte proliferation and reepithelialization [33]. These authors also used a skin organ culture system to show that the defective wound closure seen with TCRγ$^{-/-}$ mouse skin could be corrected by adding either FGF-7 or activated DETCs, with kinetics suggesting that FGFs may be sufficient to explain the stimulatory effect of DETCs on wound healing.

FGF-7 and 10 are not the only growth factors that can be produced by activated γδ-T cells. Freshly isolated human γδ-T lymphocytes, but not αβ-T cells, have also been shown after cytokine stimulation to secrete connective tissue growth factor (CTGF) [34]. A protective role for γδ-T cells during normal cell turnover in tissue homeostasis and in response to injury has been reported in other organs besides skin, including intestine, liver, and lung [35]. Moreover, γδ-T lymphocytes have been shown to interact in various ways with other cells of the immune system, including αβ-T cells, often with the effect of down-regulating inflammation [31]. Collectively, the data suggest that γδ-T cells regulate both developmental and inflammatory responses while protecting the integrity and function of their resident tissue [35].

## 2.4 "Immune" Functions of Fibroblasts

As if to underscore the extent to which the immune and repair responses are integrated in damaged tissue, fibroblasts are now also recognized as major regulators of inflammation [36, 37]. Fibroblasts, which far outnumber hematopoietic cells in all nonlymphoid organs, are the cells quintessentially involved in reconstructing tissues, and a wealth of recent evidence has emerged identifying these cells as sentinels for detecting tissue damage or infection. As the principal producers of the ECM's structural components, fibroblasts synthesize and maintain the supporting framework (stroma) of all organs and display great versatility in providing various mechanical properties in different forms of connective tissue. The cells commonly referred to as "fibroblasts" actually constitute a heterogeneous population of mature and immature cells with different embryonic origins and phenotypes in different organs and regions of the body. Immature fibroblasts have the capacity to differentiate into cells of all types of connective tissue, cartilage, bone, adipose tissue, and smooth muscle. In addition to their support role, stromal fibroblasts of bone marrow and other organs of the immune system are paracrine regulators of stem cells in the various hematopoietic and lymphoid lineages. Such effects may be important for mesenchymal stem cells of fibroblasts and other cell types, originating possibly from the bone marrow, that circulate in the blood with a potential role in tissue regeneration in various organs [38].

At least three interrelated mechanisms have been implicated in the control of local fibroblast activity during inflammation and the onset of tissue regeneration. First, the cytokine profiles produced by subsets of T-helper cells can also be synthesized at injury sites by fibroblasts, macrophages, and mast cells, with positive and negative feedback interactions again tending to polarize either the Th1 or Th2 cytokine profile [30]. Fibroblasts themselves are also target cells for many of these factors and the nature of the local cytokine profile therefore affects the rate of tissue repair and the degree of fibrosis. Sime and O'Reilly [30] suggest that Th1 cytokines generally promote regeneration of normal tissue architecture, while the Th2 response favors fibroblast activation, collagen production, and fibrogenesis. The nature, concentrations, and persistence of the cytokines in the local microenvironment are therefore important determinants of the regenerative response to injury and whether or not fibrosis occurs.

A more well-studied mechanism regulating fibroblast activity in tissue repair involves transforming growth factor-β (TGF-β), synthesized by many cells including fibroblasts, certain epithelial cells, macrophages, and T lymphocytes. TGF-β has three isoforms with production and activity regulated at many levels, from transcription to association with binding proteins in the ECM. TGF-β stimulates fibroblast proliferation, upregulates production of collagen and other ECM proteins, and is directly fibrogenic in vivo [30]. IFN-γ, the Th1 cytokine that suppresses fibroblast activity and growth, also inhibits synthesis of TGF-β, and conversely TGF-β inhibits the activity of IFN-γ [30]. The antagonism between these two factors may allow fine tuning of the balance between tissue repair and fibrosis and again emphasizes the importance of the local cytokine concentrations and activities in injured tissues.

Finally, like APCs and B cells, fibroblasts and certain other structural cells express the surface receptor CD40 and are therefore able to receive an additional activation signal from the CD40 ligand (CD40L) on the surface of T-helper cells, mast cells, basophils, eosinophils, and platelets [36]. Fibroblasts and other CD40-expressing cells in an acutely injured tissue can therefore be additionally activated by these extravasated immune cells. The CD40L signaling pathway activates the nuclear factor-κB (NF-κB) transcription factor which in fibroblasts stimulates synthesis of additional cytokines, ECM components, and cyclooxygenase-2 (COX-2) [37]. COX-2 is the inducible version of the enzyme catalyzing the rate-limiting steps in prostanoid synthesis, which in fibroblasts yields primarily prostaglandin $E_2$ ($PGE_2$). With a complex set of targets, $PGE_2$ mediates several aspects of inflammation, including both pain and fever in mammals [39]. $PGE_2$ also stimulates production of Th2 cytokines while inhibiting synthesis of Th1 cytokines, thus promoting the profibrogenic milieu discussed above. The presence of $CD40^+$ fibroblasts in a wound is clearly another mechanism promoting fibrotic activity, with likely negative effects on the tissue's capacity for regeneration.

## 2.5 The Cutaneous Immune System

The importance of defensins and mitogens derived from epidermal cells, together with the cutaneous distribution of the immune cells described briefly here, show skin to be an extremely important peripheral outpost of the immune system [40]. As shown in Fig. 2, epidermis contains LCs and DETCs, both of which can secrete cytokines that are mitogenic for keratinocytes of the

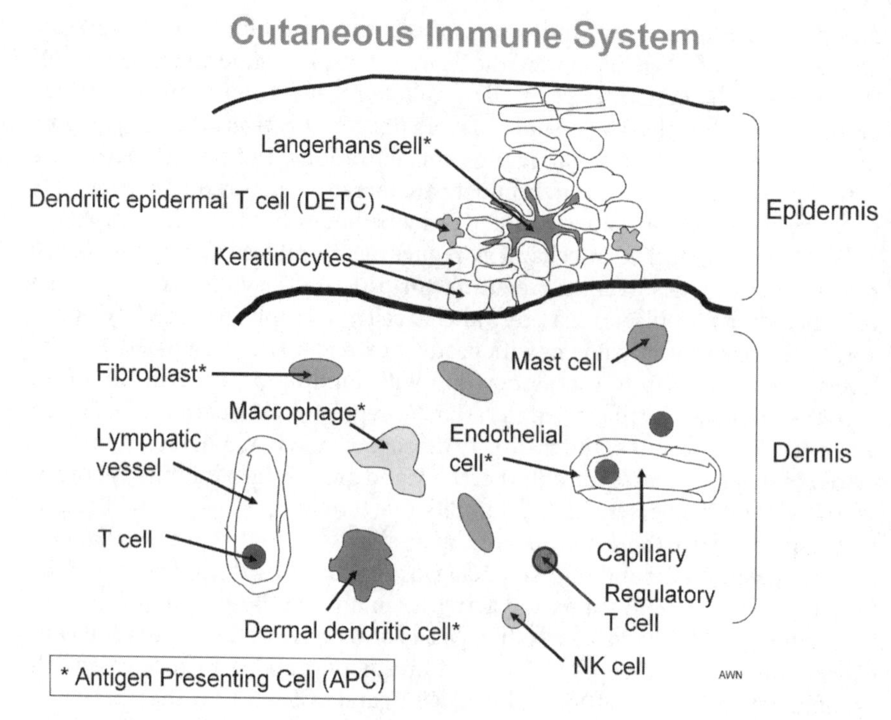


**Fig. 2** Summary of the various cells within skin which have been shown to be part the immune defense function. Plasma cells and granulocytes could also be included

epidermis and other cells. Connective tissue of the underlying dermal and hypodermal layers normally has a significant, motile population of DCs, macrophages, mast cells, NK cells, and lymphocytes, all of which can affect activities of the predominant fibroblasts in several ways; with inflammation granulocytes also appear. The number and variety of immune mechanisms that have evolved within skin is not surprising in light of that organ's role as the first line of defense against invading microbes and the requirement for almost continuous repair in some part of the organ.

## 3 A Regenerative Response in Mammals: Scar-Free Repair of Fetal Skin

### 3.1 Scarring and Inflammation

Healing of full-thickness incisional skin wounds without scarring has been documented in embryonic and early fetal stages of birds and several mammalian species [41–43]. With very little inflammation such wounds undergo rapid epithelial closure, followed by development of a normal pattern of reticular collagen and capillaries in the dermis. If injured before epidermal appendages have begun to develop, skin not only heals without fibrosis, but also develops a normal complement of hair follicles and skin glands. In contrast to repair in the adult integument, where fibrosis and wound contraction produce scars lacking normal organ structure and function, embryonic skin repair has been considered a rare example of true regeneration in mammals [2]. As such the system has become a focus of considerable interest for regeneration biologists as well as surgeons.

Scarless wound healing occurs throughout embryonic development and during fetal stages equivalent to the early second trimester of human gestation, after which there is a transition to the typical scarring repair seen with adult skin. The basis of scarless repair is multileveled and poorly understood, but considerable evidence points to a central role for components of the immune system. The period of scar-free skin healing precedes development of platelet-forming cells and most other cells of the immune system [44]. Even when myelopoiesis is underway, circulating leukocytes appear unable initially to be elicited and accumulate at fetal wounds [44]. The acute inflammatory response is minimal in fetal skin regeneration, but during the transition to adult-type scarring there is a gradual increase in the level of inflammation and specific changes in the types of immune cells elicited in the wound [45].

The sterile aqueous environment within the amniotic cavity is not required for the regenerative response of fetal skin since similar scar-free healing also occurs in newborn marsupials such as the opossum, whose development approximates that of a three month human embryo [46]. Even after two days in the mother's pouch, wounds of newborn opossum skin heal perfectly with virtually no inflammation and proportionately more monocytes/macrophages than neutrophils [45]. The dispensability of the fetal environment for scarless wound healing is also shown by regeneration of fetal skin in organ culture following incision injury and after transplantation to adult nude mice [45].

The regenerative response in fetuses is however dependent on both the type of injury and the organ involved. More severe injury to skin, producing localized cellular necrosis, evokes more typical inflammation, with large numbers of macrophages immigrating to the site, and a scar subsequently develops [41, 47]. Such experiments demonstrate that fetal monocytes/macrophages do have the capacity to respond to chemotactic signals, accumulate in wounds and mount an inflammatory response, and that fetal dermis is in fact capable of scarring. It is also important to note that the fetal mammalian regenerative response is organ specific: even at early times when skin regenerates, similar incisional wounds to the heart, diaphragm, trachea, stomach, or intestine heal with scars [41, 45, 48, 49]. Like the transplantation study mentioned above [45], these experiments suggest the importance of local rather than systemic factors in allowing skin regeneration.

## 3.2 Regenerative Potential, Immunity and the Extracellular Matrix

The dermal ECM of embryonic or early fetal skin differs in many respects from that of fully developed skin, not only histologically and in terms of its structural proteins, but also in its content of growth factors, immune cells, and other components affecting local cellular activities. Many of these differences are likely to affect the response of the skin to incisional injury and some of them may do so through immunomodulatory mechanisms. The defining feature of the ECM of embryonic mesenchyme and the developing dermis of fetal skin is high molecular weight hyaluronic acid (HA, also called hyaluronate and hyaluronan). HA is a nonsulfated glycosaminoglycan consisting of the disaccharide *N*-acetylglucosamine-glucuronate, regularly repeated up to 25,000 times in a very long, unbranched polymer. Synthesized directly into the extracellular compartment by an enzyme complex in the cell membrane, the hydrophilic HA causes a high content of water in mesenchyme and produces other features of immature connective tissue, and this has a major effect on initial events in the fetal skin wound microenvironment [50].

Chen and Abatangelo [50] have reviewed evidence that during wound repair HA directly promotes fibroblast migration, helps protect these and other cells from free radical damage, helps regulate both the amount and nature of collagen produced by fibroblasts, and binds various inhibitors of serine proteinases such as plasmin, cathepsin G, and MMP activators. While HA synthesis occurs in both adult and fetal skin wounds, it persists longer in the fetal situation and this may moderate the inflammatory response in several ways to reduce scarring [50].

Experimental removal of HA at fetal wound sites causes excessive collagen deposition and healing more like that of adults [51]. High molecular weight HA, in contrast to its degradation fragments, inhibits both leukocyte infiltration and angiogenesis in injured tissues [52, 53]. Thus the HA-rich microenvironment of fetal wounds may account at least in part for the much lower accumulation of macrophages and other inflammatory cells compared to the granulation tissue of adult skin wounds.

Another ECM component that accumulates more rapidly and at higher concentrations in fetal than in adult skin wounds is tenascin-C, a large asterisk-shaped complex of six similar glycoprotein subunits [54]. Like thrombospondins, tenascins comprise a structurally diverse group of "de-adhesive" ECM components which produce an "intermediate" type of cell adhesion. As substrates these factors readily allow cell shape changes and motility, without either the apoptotic response associated with weak substrate adhesion or the differentiative response associated in some cells with formation of focal adhesions [55]. The presence of tenascin-C in the ECM actually antagonizes the pro-adhesive effects of collagen, laminin, and other components, a property proposed to help maintain the undifferentiated state. Tenascin-C also has been shown to modulate directly the activation of T-cells and the types of interleukins produced [56, 57]. The rapid accumulation of HA, tenascin, and other matrix components in fetal skin wounds may thus help suppress the pro-inflammatory effects of any cytokines released after injury and inhibit fibrosis during healing, thereby promoting regeneration of the normal organ [41].

## 4
## Regeneration in Anuran Amphibians

### 4.1
### Regeneration of Developing Larval Limbs

Tissue and organ regeneration is widespread in amphibians, occurring in limbs, tails, jaws, and eye components (e.g., lens and neural retina) of both urodeles (salamanders and newts) and larval anurans (frogs and toads) [4, 58]. In most of these examples regeneration involves dedifferentiation of cells at the site of injury, followed by their proliferation to produce a blastema out of which the missing tissues of the damaged organ reform. Such a blastema-based regenerative process has been termed "epimorphic regeneration" and has been most thoroughly studied in urodele limbs [6, 59]. Blastema growth involves production of both HA [60, 61] and tenascin-C [62]. The excellent capacity for epimorphic regeneration in urodeles may be related to the fact that they initiate deconstruction of the ECM and cellular dedifferentiation in response to amputation with minimal inflammation, immune cell infiltration, and scarring [58].

The gradual loss of regenerative ability in developing tissues and organs such as hindlimbs in anurans as they approach and undergo metamorphosis has been observed in many species and studied most thoroughly in *Xenopus laevis* [11, 63]. Hindlimb buds first appear at stage 47 [64] and regenerate perfectly if partially removed at any time through stage 52, which is approximately when chondrogenesis of the femur begins. The regeneration process, like that of urodele limbs, involves rapid epithelial closure of the amputation wound, thickening of the resulting wound epithelium, dedifferentiation and accumulation of proliferating mesenchymal cells under the influence of the wound epithelium, followed by distal outgrowth of the blastema and re-differentiation of new limb tissues in continuity with those of the stump [59]. Inflammation and fibrosis in the regenerating limb are minimal [58]. Filoni and colleagues have clearly shown that, unlike limb regeneration in urodeles, the process in the *early* developing anuran hindlimb does not depend on the organ's nerve supply [65].

Amputation of developing anuran hindlimbs at progressively later stages, during the "prometamorphic period" [66], results in increasingly deficient regeneration manifested by production of smaller limbs with less muscle and fewer digits [63]. Regenerative capacity is lost first in proximal limb regions. At any given stage of metamorphosis, distal amputation of a hindlimb results in more complete regeneration than a proximal amputation [67]. Since limb tissues develop in a proximal-to-distal direction, it has been assumed that regenerative potential is lost in general as tissues differentiate. The data of Wolfe [63] are consistent with previous work, but also show that regenerative capacity is greater at any stage when amputation is through a joint region. This was particularly apparent at later stages of prometamorphosis with limbs amputated through the ankle or tarsal-metatarsal joint rather than through the tarsus. These authors suggested that the state of bone ossification at the amputation site is important for the regenerative outcome, but Harty et al. [11] raise the possibility that the ECM of the developing joint may provide conditions more conducive to epimorphic regeneration. Both HA and tenascin-C are highly concentrated in the differentiating connective tissues of developing joints [68, 69] and the resulting properties of the local ECM may contribute to the greater regenerative capacity of these regions.

If amputated during late prometamorphosis or as young froglets *Xenopus* hindlimbs usually produce a heteromorphic, non-segmented "spike" of skin-covered cartilage. At these stages "fibroblast-like" cells in the limb stumps aggregate beneath the wound epithelia as "pseudoblastemas," which contain much collagen and appear to be derived from connective tissue [70, 71]. Like blastema cells of urodele limbs, growth in pseudoblastemas depends both on nerves [72, 73] and the wound epithelium [74]. Most evidence is consistent with the view that the growth of spikes represents abortive or pattern-deficient epimorphic regeneration rather than simple tissue repair or chondroplasia [73, 74].

Expression of various genes involved both in limb formation and regeneration occurs in post-metamorphic *Xenopus* forelimbs that form only pseudoblastemas and hypomorphic spikes. These include two negative regulators of tissue differentiation, *Id2* and *Id3* [75], *HoxA13*, *msx1*, and *fgf*-8 [73], and a regulator of dorsal-ventral patterning, *Lmx*-1 [76]. Histology reveals that tissue dedifferentiation in the limb stump is not extensive and that redifferentiation, particularly of cartilage, occurs very quickly [71]. High mitotic activity persists in the most distal region of the stump, giving rise to the cartilaginous spike [73]. Expression of both *Id3* and *Lmx-1* is weak and rapidly downregulated [75, 76]. Such studies suggest that distal cells of regeneration-incompetent, spike-forming limb stumps do undergo limited dedifferentiation and expression of genes required for regeneration, but that these processes are squelched so that formation of a normal blastema never occurs.

It has been suggested by several studies that the loss of regenerative potential during anuran development involves fibrosis, with proliferation of fibroblast-like cells inhibiting or interfering with the normal course of blastema formation and limb regeneration [70, 77, 78]. Regeneration in limbs of juvenile frogs can be improved by various nonspecific treatments to stimulate stump tissue hydrolysis and dedifferentiation or to interfere with fibrosis at the wound site [79]. Yokoyama et al. [80] reported that stage 56 *Xenopus* hindlimbs, which have lost considerable regenerative capacity, regenerate almost normally following implantation of beads containing FGF-10 in the distal limb stumps. These studies indicate that the lost regenerative capacity of anuran limbs can be restored and that factors present locally in the stump tissues are more important than changing hormonal or other systemic conditions. Work on the regenerative ability of *Xenopus* limbs reciprocally transplanted between larval and post-metamorphic stages also underscores the importance of local rather than systemic changes [81].

It should be noted that the potential for scarless wound healing in anurans was directly tested by Yannas et al. [82] who examined repair of full-thickness excisional skin wounds on larval and adult frogs (*Rana catesbiana*). The capacity to regenerate normal skin was greatest in young larvae but declined continuously during development, while contraction became the dominant mode of wound closure. The shift from regeneration to contraction in closure of skin wounds was accelerated at late prometamorphic stages. Scarring was found only in wounds of postmetamorphic animals [83]. This work indicates that fundamental changes during larval development cause diminished capacities for both epimorphic limb regeneration and scar-free skin regeneration.

## 4.2 Proposed Causes of Regenerative Failure in Adults

Improved understanding of the changes in anuran skin and hindlimbs related to metamorphosis would strengthen greatly the theoretical basis of attempts to improve regeneration in mammals. Various reasons for the regenerative failure of limbs have been suggested by work on amphibian regeneration. From studying the trophic influence of nerves in urodele limb regeneration, Singer [84, 85] favored the view that the declining regenerative capacity of developing anurans is due to a reduced quantity of innervation available for cells at the level of amputation. This idea is consistent with the observation that scar-free wound healing in fetal skin requires an adequate nerve supply [86]. However quantitative analysis of the nerve supply in limbs of a wider range of regenerating and nonregenerating amphibian species failed to support the nerve-limitation hypothesis [87]. This view was further undermined by the studies demonstrating that regeneration of developing *Xenopus* hindlimbs does not require nerves at all, apparently because the factors supplied by nerves are available locally from other cells in the developing tissues [65].

The problem of diminishing regenerative capacity during anuran development was also addressed by Schotté [88] in a review of work on endocrine ablation and hormonal changes during metamorphosis. These studies showed that neither thyroid hormone nor any pituitary hormone are directly required for limb regeneration but that corticosteroids play critical roles in determining the extent of regeneration vs fibrosis following limb amputation [88, 89]. This is consistent with the well-known effects of adrenal corticoids such as cortisone on inflammation and wound healing. The endocrine studies again emphasized the importance of *local* rather than systemic factors in determining an organ's capacity to regenerate.

Another explanation adduced for regenerative failure in anuran limbs is that the wound epithelia on such limb stumps do not become functional, i.e., do not maintain proliferation of underlying mesenchymal cells [90]. The potential importance of this view was emphasized by work showing expression of *fgf* in wound epithelia of amputated *Xenopus* hindlimbs at regeneration-competent, but not regeneration-incompetent stages [65, 91]. Although many alterations in anuran skin during metamorphosis have been described [66], little is known of the changing physiological or developmental interactions in the epidermis or other skin elements that might affect properties such as the apparent loss of *fgf* expression in wound epithelia.

## 4.3 Changes in the Immune System and Regenerative Potential

Among the many metamorphic changes in amphibian organ systems, those involving cells of the immune system are of particular interest here because of their roles in tissue repair and because they constitute an intrinsic part of specific organs such as skin [33]. As we have seen such cells are of primary importance in determining how a tissue responds to injury, the degree of inflammation, and the extent of scarring [28, 92, 93]. Perhaps because so little was known about these aspects of the amphibian immune system until recently, there have been relatively few investigations of immune cell function during limb regeneration [94]. As described above, current studies in several laboratories implicate immune cells and their regulators in the regenerative aspects of fetal wound healing.

Based primarily on studies with *Xenopus* considerable evidence has accumulated indicating that both the cellular and humoral immune responses of adult anurans are remarkably similar to those of mammals in terms of specificity, speed of onset, and memory [95, 96]. Skin of anurans contains both Langerhans cells and DCs, as demonstrated by histochemistry [97], as well as MHC class II positive DCs [98]. The phylogenetically more primitive urodele amphibians, though not as well studied, appear to be "immunodeficient" compared to adult frogs and toads, with a minimal inflammatory response, relatively weak cellular immunity, and humoral immunity that is slow, produces only IgM and lacks memory cells [99].

Interestingly in light of their regenerative capacity, larval anurans' immune responses appear to be much less well-developed and efficient than those of their postmetamorphic counterparts [96]. As measured by allogenic skin graft rejection, cellular immunity of larvae is both weaker and slower. Larval antibodies against a standard antigen are less diverse and of lower affinity than those of genetically identical adults. Thus both the T-cell and B-cell recognition repertoires of tadpoles appear to be more restricted than the adult repertoires [96]. Cells expressing MHC class I and II proteins, including APCs of skin, are also much more restricted in larvae than in adults. Finally, other elements of the innate immune response are also less well developed in larvae: antimicrobial peptides of the skin do not develop until metamorphosis and NK cell activity toward allogenic tumor cells is present in adults but not tadpoles [96].

Clearly, significant refinements in immunity and major turnover of lymphocytes and other immune cells occur in the anuran during the transition from larva to adult. Some memory cells of both humoral and cellular immunity are retained through metamorphosis [96], but there is evidence for self/non-self histoincompatibility between larval and adult tissues. For example Izutsu and Yoshizato [100] reported rejection of *Xenopus* skin removed from larvae, cryopreserved and then autografted back onto the donors after metamorphosis. During the metamorphic transition itself the anuran immune response is temporarily suppressed, possibly due to lymphocyte apoptosis induced by transiently elevated corticosteroid levels [96, 101]. Nearly every organ and tissue is being remodeled during this time and the temporary inhibition of lymphocyte growth and function may allow the metamorphic organism to adjust its perception of 'self' and subsequently re-establish tolerance to new self-antigens when adult cells predominate in these organs [96, 102]. Rollins-Smith [96] suggested that the reorganization of the anuran immune system may serve to eliminate unnecessary lymphocytes that would be potentially destructive if they recognized new adult-specific antigens in remodeled tissues.

## 5 Interrelationship of Immunity, Regenerative Capacity, and Scarring

### 5.1 Complement, TLRs, and Mitogens from γδ-T Cells

As indicated earlier many cells and factors specialized for immune protection have been found to play additional roles in growth and development. Components of the innate immune system, including the antimicrobial polypeptides secreted by epithelial cells and by phagocytic cells summoned to sites of tissue injury or invasion by microorganisms, stimulate cell proliferation or survival and thus directly help maintain the integrity of the organism [14, 15]. Factors generated as part of the complement system for bacterial cell lysis, specifically C3a and C5a, are also produced locally during endochondral ossification [16], in liver regeneration after partial hepatectomy [103], and in both lens and limb regeneration in newts [104]. In addition the urodele ortholog of CD59, a major regulator of complement activation, has been shown to be involved in mediating positional information during limb regeneration [105].

Both NK cells and macrophages release MMPs and growth factors and thereby play important roles in tissue repair and regeneration by affecting cellular interactions with the ECM and rates of proliferation [21]. Though clearly important for wound debridement, macrophages appear not to be critical for healing since skin wounds in mice genetically unable to produce such cells heal efficiently, possibly because comparable MMPs and cytokines are produced by other, nonmyeloid cells [23]. Recent work suggests that any cell of the innate immune system expressing a full complement of TLRs may secrete proteolytic, mitogenic, and angiogenic factors when it binds various bacterial components or "heat shock proteins" released from injured cells [27].

Similar stress proteins also elicit expression of various polypeptide growth factors, including FGFs, from γδ-T cells. Transgenic mice lacking such cells show significantly inhibited repair of skin wounds owing to the lack of FGF-7 expression in the injured tissue [33]. Other factors released by γδ-T cells produce a local inhibition of inflammation and are likely also to be important for tissue regeneration [31].

## 5.2 Cytokine Levels in the Balance Between Regeneration and Fibrosis

Fibroblasts express various receptors once thought to be found strictly in cells of the immune system. Proliferation of fibroblasts and collagen production are regulated by various interleukins synthesized by immune cells during inflammation and early repair, including CD40L, which also induce fibroblast secretion of other inflammatory mediators [30].

Considerable evidence suggests that components of the immune response are responsible for the loss of true regenerative ability in wounded skin, i.e., for scarring. As we have seen, certain immune cells, including DCs, DETCs, and fibroblasts themselves, are normally abundant in skin and are thus among the candidates for regulating scarless wound healing [106]. Activated fetal fibroblasts have been shown to secrete significantly less of the pro-inflammatory cytokines IL-6 and IL-8 than fibroblasts from adults and in vivo fetal skin produces minimal levels of IL-6 after injury compared to adult skin [107]. IL-6, like IL-8, recruits monocytes to injured tissues and also directly activates these cells to form macrophages. Therefore, less production of these cytokines in fetal skin may explain the decreased number of macrophages and the minimal inflammatory response during scarless wound healing.

IL-10, although part of the Th2 cytokine profile which Sime and O'Reilly [30] suggest promotes fibrosis, is an anti-inflammatory cytokine that inhibits expression of both IL-6 and IL-8 and deactivates macrophages [108]. The balance of these cytokines in a wound may thus also help determine the resulting degree of inflammation and scarring, an idea supported by work with experimentally manipulated IL-10 production in murine skin. Skin from IL-10 null mice at embryonic day 15 was grafted to same strain adult mice and incisionally wounded five days later. After one week the wounds contained significantly greater densities of inflammatory leukocytes than day 15 fetal skin from wild-type mice grafted and wounded in the same manner [109]. Moreover, unlike the fetal skin controls, that from the IL-10$^{-/-}$ mice produced definite scars by one week after wounding [109]. Conversely, IL-10 *overexpression* in skin of adult mice was reported to decrease inflammation after wounding and to allow scar-free, regenerative healing [110]. By regulating production of other cytokines, IL-10 may be a major factor promoting regeneration rather than inflammation and fibrosis.

Studies with cytokines in fetal skin wounds suggest strongly that suppressed inflammation and reduction in the number of macrophages and other leukocytes produces less scarring and better regeneration. The recent work of Martin et al. [23] with PU.1 null mice, indicating that neither neutrophils nor macrophages are required for normal repair of skin wounds, provides further evidence for this view. Adult PU.1 null mice, lacking both these cell types, were found not only to heal skin wounds as rapidly as controls despite their lack of these inflammatory cells, but also to do so in a scar-free manner (Fig. 1). Like fetal skin wounds those of adult PU.1 null mice expressed almost no mRNA for IL-6, which is abundant in adult skin wounds. These authors suggest that inflammatory cytokines are not an essential prerequisite for tissue repair, at least in the absence of infection, and are responsible for fibrosis and scarring. It should be noted that the PU.1 null mice also lack myeloid DCs, but not lymphoid DCs [24], indicating that these cells as well as macrophages and neutrophils are possible sources for profibrotic cytokines interfering with fetal skin regeneration.

The general view that at least some aspects of inflammation cause events leading to fibrosis is consistent with additional studies showing that topical treatment of wounds in adult mice with anti-inflammatory COX-2 inhibitors inhibits several parameters of inflammation locally and results in significantly reduced production of scar tissue [111].

The cytokine most directly implicated in fibrotic activity is TGF-β, which is released from degranulating platelets in the immediate aftermath of tissue injury. During repair TGF-β remains available, sequestered and regulated by binding proteins, and various isoforms are also synthesized locally by macrophages and other cells, including fibroblasts and keratinocytes [2]. Developmental control of TGF-β activity occurs at many levels, all of which seem to be involved during tissue repair [48, 112]. Such controls include complex regulation of transcription and differential expression of TGF-β isoforms, various transmembrane receptors and extracellular binding proteins and proteoglycans. Compared to adult skin wounds, levels of TGF-β mRNA are reduced in fetal wounds, the extent varying with gestational age and the availability of macrophages. Interestingly, TGF-β mRNA levels are greatly diminished in the macrophage-free skin wounds of PU.1 null mice [23].

TGF-β activity is clearly important in skin wound repair. Application of exogenous TGF-β1 to fetal skin wounds induces scarring [41], while reducing TGF- β1 and 2 activity in adult skin wounds by antibodies, antisense oligonucleotides, or a synthetic antagonist inhibits fibrosis and scar formation [113, 114]. The wound studies are consistent with a large body of other literature demonstrating the importance of TGF-β in various medical conditions and disease states involving fibrosis, all of which suggest a central role for this cytokine in the molecular events underlying repair and scarring. Multiple, interrelated roles for TGF-β in tissue repair are indicated by work of

Roberts [115] who examined healing of skin wounds in mice null for Smad3, a signal transduction factor for TGF-β. Not only was the fibrotic response reduced in the $Smad3^{-/-}$ mice, as expected from other studies, but wound reepithelialization and the overall rate of healing were greatly accelerated compared with wild-type mice.

## 5.3 Adaptive Immunity and the Inhibition of Regeneration

In contrast to mammalian wounds, very few studies of amphibian limb regeneration have focused directly on events of inflammation. There have been no analyses of specific interleukins or immune cells present in the amphibian limb stump during tissue dedifferentiation or blastema growth. As indicated here, however, comparative immunologists have shown that the immune mechanisms of adult anurans resemble those of phylogenically more advanced vertebrates in most respects. Based on this work we expect that similar cytokines and similar balances between humoral and cellular immunity are likely to be operative in both postmetamorphic anuran limb stumps and mammalian wounds.

We view the gradual loss of limb regenerative ability in the frog as it approaches and completes metamorphosis as analogous to the transition in the developing mammal from fetal (scar-free) to adult (scarring) skin repair. In both situations many variables are changing steadily, including the composition of the ECM, cells in the skin, and the "differentiated state" of the tissues, but as indicated here changes in the cells and signaling factors of the immune system are likely to be of major importance. The demonstration of greatly reduced scarring and more perfect skin regeneration in adult mice with specific immune deficits following knockout of various cytokines or immune cells [23, 115] is strong evidence for a role of immune components in the loss of regenerative ability in skin. Similarly the "immunodeficient" state of urodele amphibians compared to adult *Xenopus* may help explain their lifelong greater capacity for organ regeneration.

How might the more highly developed immune system act to block an organ's capacity for regeneration after injury? Cells undergoing dedifferentiation early in regeneration may, like some virus-infected cells or neoplastic cells, be recognized and eliminated by cytotoxic T cells or NK cells if they display antigens recognized in the adult as "non-self." Cell-mediated killing of dedifferentiating cells could allow subsequent excessive growth and activity of fibroblasts or other cells that might further compromise organ regeneration. There is no evidence for removal of either dedifferentiated cells or stem cells in nonregenerating animals, possibly because the hypothesis has never been tested.

An alternative to cell removal is the possibility that the cellular interactions required to initiate regenerative processes, such as development of epidermal appendages or growth and patterning of a limb blastema, are simply blocked by excessive fibroplasia and collagen production driven by cytokines released during inflammation [2, 58]. Wounds or other tissue insults involving a prolonged or intensified inflammatory response invariably heal with fibrosis or contraction and as we have seen scarring in wounds can be reduced dramatically by experimentally eliminating one or more cytokines released during inflammation [113, 114].

Just as the adaptive immune system apparently arose in primitive vertebrate predators as part of the evolutionary response to more injuries and infections [12], the inflammatory response became more highly developed in animals that were able to leave their aqueous environments [116]. Under such conditions the advantage lay with those animals able to close wounds rapidly with tough, impermeable scar tissue to minimize infection, fluid loss, and further tissue damage. Selective pressure favoring rapid scar production likely increased when vertebrates became homeothermic and wounds became warm, moist and nutrient-rich niches in which bacteria and other microorganisms could escape cold, less nutritive environments [116]. With regard to mammalian skin wounds, Ferguson et al. [117] have argued that evolution produced an inflammatory and scarring response that overrides the endogenous regenerative potential of the organ seen in more primitive vertebrates. The capacity to regenerate and restore complete biological function has been superseded by mechanisms that produce rapid wound closure and fibrotic repair [118].

Goss [1, 8] postulated that regenerative ability in larger appendages such as limbs was selected against during vertebrate evolution, especially after the appearance of homeothermy, but could not speculate on the mechanisms involved. We suggest that the advantages of epimorphic regeneration came to be outweighed by refinements in the adaptive immune system, especially those that enhanced the acute inflammatory response in wounds [11]. Such immune mechanisms not only allowed an accelerated and specific defense against microbial invaders, but also resulted in rapid growth of a thickened protective barrier at the wounded site.

In humans the pro-fibrotic inflammatory response which we propose inhibits regeneration produces many examples of pathological scarring diseases, such as the autoimmune disorders scleroderma and rheumatoid arthritis and the pulmonary fibrosis caused by prolonged environmental exposure to certain antigenic agents [30, 118]. In such fibrotic diseases we see that the phylogenetically modern inflammatory response can harbor other disadvantages besides preemption of the regenerative response.

## 6 Summary and Conclusions

The protective functions of the immune system are intimately associated with mechanisms for tissue maintenance and functional restoration after injury. From the antimicrobial polypeptides and complement system of primitive vertebrates to the highly specific lymphocytes of mammals, factors specialized for defense against microorganisms are concomitantly involved with localized stimulation of growth in the affected tissues. Moreover, recent work indicates that fibroblasts, like lymphocytes and other immune cells, both synthesize and respond to various cytokines during inflammation. In wounds of vertebrates adaptive immunity usually produces fibroplasia, contraction and collagen production resulting in a scar.

Scarring is incompatible with tissue and organ regeneration. Full-thickness incision injuries to fetal skin, which are repaired with minimal inflammation and without scarring, represent an increasingly well-characterized model for vertebrate organ regeneration. Loss of the ability to heal incisions in a scar-free manner occurs gradually during fetal development in conjunction with formation of the immune system and the onset of the inflammatory response to injury. Studies with specific growth factors, especially TGF-β, and with various transgenic mouse models deficient for specific immune cells or interleukins reveal the importance of inflammatory cytokines in the fibrotic reaction and inhibition of complete skin regeneration.

We propose that the failure of regeneration in limbs and other organs of adult anurans, reptiles, and phylogenetically more advanced vertebrates is also due to the status of the cells and signaling systems for adaptive immunity in these organisms. Vertebrates with less well-developed adaptive immunity, such as teleost fish, larval anurans, and urodele amphibians throughout life, show minimal inflammatory responses to injury and excellent capacities for tissue and organ regeneration. During the premetamorphic period of anurans, hindlimb amputation produces increasingly less regeneration of distal limb structures and more fibrosis or chondroplasia. In *Xenopus* larvae this transition coincides with formation of a new immune system more like that of mammals.

As work with scarless wound healing has already indicated, important insights into the loss of regenerative ability during the course of evolution are likely to come from improved understanding of the cellular activities and factors released during inflammation that cause fibrosis and wound contraction. Echoing the assertion of Goss [1] with which we opened this chapter, such new understanding will be necessary before intelligent attempts can be made to restore regeneration where it does not normally occur.

**Acknowledgments** Research in our laboratories was funded by grants from NSF and the Indiana 21st Century Research and Technology Fund. We wish to thank Elizabeth and Ron Osborne for help with the manuscript.

## References

1. Goss RJ (ed) (1969) Principles of regeneration. Academic Press, New York
2. Martin P (1997) Wound healing: aiming for perfect skin regeneration. Science 276:75
3. Thouveny Y, Tassava RA (1997) Regeneration through phylogenesis. In: Ferretti P, Geraudie J (eds) Cellular and molecular basis of regeneration: from vertebrates to humans. Wiley, Chichester UK, p 9
4. Tsonis PA (ed) (1996) Limb regeneration. Cambridge University Press, New York

5. Tsonis PA (2000) Dev Biol 221:273
6. Tsonis PA – this volume
7. Ferretti P, Geraudie J (eds) (1998) Cellular and molecular basis of regeneration: from invertebrates to humans. Wiley, New York
8. Goss RJ (1992) J Theor Biol 159:24
9. Alvarado AS (2000) BioEssays 22:578
10. Brockes JP, Kumar A, Velloso CP (2001) J Anat 199:3
11. Harty M, Neff AW, King MW, Mescher AL (2003) Dev Dyn 226:268
12. Matsunaga T, Rahman A (1998) Immunol Rev 166:177
13. Hoffmann JA, Kafatos FC, Janeway CA Jr, Ezekowitz RAB (1999) Science 284:1313
14. Elsbach P (2003) J Clin Invest 111:1643
15. Sorensen OE, Cowland JB, Theilgaard-Monch K, Liu L, Ganz T, Borregaard N (2003) J Immunol 170:5583
16. Mastellos D, Lambris J (2002) Trends Immunol 23:485
17. Egozi EI, Ferreira AM, Burns AL, Gamelli MD, Dipietro LA (2003) Wound Rep Regen 11:46
18. Singer AJ, Clark RAF (1999) New Eng J Med 341:738
19. Simpson DM, Ross R (1972) J Clin Invest 51:2009
20. Pardoll DM (2001) Science 294:534
21. Moretta A, Bottino C, Mingari MC, Biassoni R, Moretta L (2002) Nat Immunol 3:6
22. Leibovich SJ, Ross R (1975) Am J Path 78:71
23. Martin P, D'Souza D, Martin J, Grose R, Cooper L, Maki R, McKercher SR (2003) Curr Biol 13:1122
24. Guerriero A, Langmuir PB, Spain LM, Scott EW (2000) Blood 95:879
25. Lutz MB, Schuler G (2002) Trends Immunol 23:445
26. Nevo U, Kipnis J, Golding I, Shaked I, Neumann A, Akselrod S, Schwartz M (2003) Trends Mol Med 9:88
27. Li M, Carpio DF, Zheng Y, Bruzzo P, Singh V, Ouaaz F, Medzhitov RM, Beg AA (2001) J Immunol 166:7128
28. Barbul A (1992) Role of the immune system. In: Cohen IK, Diegelmann RF, Lindblad WJ (eds) Wound healing: biochemical and clinical aspects. Saunders, Philadelphia
29. Naldini A, Pucci A, Bernini C, Carraro F (2003) Curr Pharm Design 9:511
30. Sime PJ, O'Reilly MA (2001) Clin Immunol 99:308
31. Hayday AC (2000) Ann Rev Immunol 18:975
32. Hayday A, Tigelaar R (2003) Nat Rev Immunol 3:233
33. Jameson J, Ugarte K, Chen N, Yachi P, Fuchs E, Boismenu R, Havran WL (2002) Science 296:747
34. Workalemahu F, Foerester M, Kroegel C, Braun RK (2003) J Immunol 170:153
35. Born WK, O'Brien RL (2002) Nat Med 8:560
36. Smith RS, Smith TJ, Blieden TM, Phipps RP (1997) Am J Path 151:317
37. Buckley CD, Pilling D, Lord JM, Akbar AN, Scheel-Toellner D, Salmon M (2001) Trends Immunol 22:199
38. Stocum DL – this volume
39. Smith WL, DeWitt DL, Faravito RM (2000) Ann Rev Biochem 69:145
40. Cruz PG Jr (2001) The epidermis: an outpost of the immune system. In: Freinker RK (ed) Biology of the skin. Parthenon, New York, p 255/chap 17
41. McCallion RL, Ferguson MWJ (1996) Fetal wound healing and the development of anti-scarring therapies for adult wound healing. In: Clark RAF (ed) The molecular biology of wound repair. Plenum Press, New York, p 561
42. Garg HG, Longaker MT (eds) (2000) Scarless wound healing. Marcel Dekker, New York

43. Lorenz HP, Longaker MT – this volume
44. Nodder S, Martin P (1997) Anat Embryol 195:215
45. Chin GS, Stelnicki EJ, Gittes GK, Longaker MT (2000) Characteristics of fetal wound repair. In: Garg HG, Longaker MT (eds) Scarless would healing. Marcel Dekker, New York, p 239
46. Armstrong JR, Ferguson MW (1995) Dev Biol 169:242
47. Hopkinson-Woolley J, Hughes D, Gordon S, Martin P (1994) J Cell Sci 107:1159
48. Stelnicki EJ, Chin GS, Gittes GK, Longaker MT (1999) Sem Ped Surg 8:124
49. Longaker MT, Peled ZM, Chang J, Krummel TM (2001) Surgery 130:785
50. Chen WYJ, Abatangelo G (1999) Wound Repair Regen 7:79
51. Mast BA, Nelson JM, Krummel TM (1992) In: Cohen IK, Diegelmann RF, Lindbad WJ (eds) Wound healing: biochemical and clinical aspects. WB Saunders, Philadelphia, p 326
52. Savani RC, Hou G, Liu P, Wang C, Simons E, Grimm PC, Stern R, Greenberg AH, DeLisser HM, Khalil N (2000) The role of hyaluronan-receptor interactions in wound repair. In: Garg H, Longaker MT (eds) Scarless wound healing. Marcel Dekker, New York, p 115
53. Balazs EA, Larsen NE (2000) Hyaluronan: aiming for perfect skin regeneration. In: Garg H, Longaker MT (eds) Scarless wound healing. Marcel Dekker, New York Basel, p 143
54. Whitby DJ, Ferguson MW (1991) Development 112:651
55. Murphy-Ullrich JE (2001) J Clin Invest 107:785
56. Hemesath TJ, Marton LS, Stefansson K (1994) J Immunol 152:5199
57. Hibino S, Kato K, Kudoh S, Yagita H, Okumura K (1998) Biochem Biophys Res Commun 250:119
58. Stocum DL (ed) (1995) Wound repair, regeneration and artificial tissues. RG Landes, Austin
59. Mescher AL (1996) Int J Dev Biol 40:785
60. Mescher AL, Cox CA (1988) Differentiation 38:161
61. Munaim SI, Mescher AL (1986) Dev Biol 11:138
62. Onda H, Poulin ML, Tassava RA, Chiu IM (1991) Dev Biol 148:219
63. Wolfe AD, Nye HL, Cameron JA (2000) Dev Biol 233:72
64. Nieuwkoop PD, Faber J (eds) (1956) Normal table of *Xenopus laevis* (Daudin). North-Holland, Amsterdam
65. Cannata SM, Bagni C, Bernardini S, Christen B, Filoni S (2001) Dev Biol 231:436
66. Shi Y-B (ed) (2000) Amphibian metamorphosis: from morphology to molecular biology. Wiley-Liss, New York
67. Dent JN (1962) J Morphol 110:61
68. Onda H, Goldhamer DJ, Tassava RA (1990) Development 108:657
69. Jarvinen TA, Jozsa L, Kannus P, Jarvinen TL, Kvist M, Hurme T, Isola J, Kalimo H, Jarvinen M (1999) J Cell Sci 112:3157
70. Komala Z (1957) Folia Biol (Krakow) 5:1
71. Korneluk RG, Liversage RA (1984) Can J Zool 62:2383
72. Korneluk RG, Anderson MJ, Liversage RA (1982) J Exp Zool 220:331
73. Endo T, Tamura K, Ide H (2000) Dev Biol 220:296
74. Goss RJ, Holt R (1992) J Exp Zool 261:451
75. Shimizu-Nishikawa K, Tazawa I, Uchiyama K, Yoshizato K (1999) Dev Growth Diff 41:731
76. Matsuda H, Yokoyama H, Endo T, Tamura K, Ide H (2001) Dev Biol 229:351
77. Goode RP (1967) J Embryol Exp Morphol 18:259
78. Carlson BM (1974) Factors controlling the initiation and cessation of early events in the regenerative process. In: Sherbert GV (ed) Neoplasia and cell differentation. Karger, Basel, p 60

79. Polezhaev LV, Pavlenko AP (1972) Sov J Dev Biol 3:313
80. Yokoyama H, Ide H, Tamura K (2001) Dev Biol 233:72
81. Sessions SK, Bryant SV (1988) J Exp Zool 247:39
82. Yannas IV, Colt J, Wai YC (1996) Wound Repair Regen 4:431
83. Yannas IV (2001) Tissue and organ regeneration in adults. Springer, Berlin Heidelberg New York
84. Singer M (1965) A theory of the trophic nervous control of amphibian limb regeneration, including a re-evaluation of quantitative nerve requirements. In: Kiortsis V, Trampusch HAL (eds) Regeneration in animals and related problems. North-Holland, Amsterdam, p 20
85. Singer M (1978) Am Zool 18:829
86. Stelnicki EJ, Doolabh V, Lee S, Levis C, Baumann FG, Longaker MT, Mackinnon S (2000) Plast Reconstr Surg 105:140
87. Scadding SR (1982) J Exp Zool 219:81
88. Schotte OE (1961) Systemic factors in initiation of regenerative processes in limbs of larval and adult amphibians. In: Rudnick D (ed) 19th Growth Symposium. Ronald Press, New York, p 161
89. Liversage RA, McLaughlin DS, McLaughlin HMG (1985) The hormonal milieu in amphibian appendage regeneration. In: Sicard RE (ed) Regulation of vertebrate limb regeneration. Oxford University Press, New York, p 54
90. Tassava RA, Olsen CL (1982) Differentiation 22:151
91. Christen B, Slack JMW (1997) Dev Biol 192:455
92. Slavin J (1996) J Path 178:5
93. Gillitzer R, Goebeler M (2001) J Leukocyte Biol 69:513
94. Sicard RE (ed) (1985) Regulation of vertebrate limb regeneration. Oxford University Press, New York
95. Robert J, Cohen N (1998) Immunol Rev 166:231
96. Rollins-Smith, LA (1998) Immunol Rev 166:221
97. Castell-Rodriguez AE, Hernandez-Penaloza A, Sampedro-Carrillo EA, Herrera-Enriquez MA, Alvarez-Perez SJ, Rondan-Zarate A (1999) Dev Comp Immunol 23:473
98. Du Pasquier L, Flajnik MF (1990) Dev Immunol 1:85
99. Tournefier A, Laurens V, Chapusot C, Ducoroy P, Padros MR, Salvadori F, Sammut B (1998) Immunol Rev 166:259
100. Izutsu Y, Yoshizato K (1993) J Exp Zool 266:163
101. Rollins-Smith LA, Cohen N (1996) Metamorphosis: an immunologically unique period in the life of the frog. In: Gilbert LI, Tata JR, Atkinson BG (eds) Metamorphosis: post-embryonic re-programming of gene expression in amphibian and insect cells. Academic Press, San Diego, p 363
102. Ruben LN, Clothier RH, Horton JD, Balls M (1989) BioEssays 10:8
103. Strey CW, Markiewski MJ, Mastellos D, Tudoran R, Spruce LA, Greenbaum LE, Lambris JD (2003) J Exp Med 198:913
104. Kimura Y, Madhavan M, Call MK, Santiago W, Tsonis PA, Lambris JD, Del Rio-Tsonis K (2003) J Immunol 170:2331
105. da Silva SM, Gates PB, Brockes JP (2002) Dev Cell 3:547
106. Kikly K, Lotze MT (2001) DCs in wound healing. In: Lotze MT, Thomson AW (eds) Dendritic cells: biology and clinical applications, 2nd edn. Academic Press, San Diego, p 539
107. Liechty KW, Adzick NS, Crombleholme TM (2000) Cytokine 12:671
108. Elgert KD, Alleva DG, Mullins DW (1998) J Leukoc Biol 64:275
109. Leichty KW, Kim HB, Adzick NS, Crombleholme TM (2000) J Ped Surg 35:866
110. Gordon AD, Karmacharya JK, Herlyn M, Crombleholme TM (2001) Surg Forum 52:568

111. Wilgus TA, Vodovotz Y, Vittadini E, Clubbs EA, Oberyszyn TM (2003) Wound Repair Regen 11:25
112. Bullard KM, Longaker MT, Lorenz HP (2003) World J Surg 27:54
113. Shah M, Rorison P, Ferguson MJW (2000) The role of transforming growth factors-beta in cutaneous scarring. In: Garg HG, Longaker MT (eds) Scarless wound healing. Marcel Dekker, New York, p 213
114. Huang JS, Wang YH, Ling TY, Chuang SS, Johnson FE, Huang SS (2002) FASEB J 16:1269 Epub 2002 June 21
115. Ashcroft GS, Yang X, Glick AB, Weinstein M, Letterio JL, Mizel DE, Anzano M, Greenwell-Wild T, Wahl SM, Deng C, Roberts AB (1999) Nat Cell Biol 1:260
116. Cooper EL (1976) Comparative immunology. Prentice Hall, Englewood Cliffs
117. Ferguson MWJ, Whitby DJ, Shah M, Armstrong J, Siebert JW, Longaker MT (1996) Plast Reconstr Surg 97:854
118. O'Leary R, Wood EJ, Guillou PJ (2002) Eur J Surg 168:523

Received: January 2004

Adv Biochem Engin/Biotechnol (2005) 93: 67–81
DOI 10.1007/b99967


# Vertebrate Limb Regeneration

Mindy K. Call · Panagiotis A. Tsonis (✉)

Laboratory of Molecular Biology, Department of Biology, University of Dayton,
Dayton, OH 45469-2320, USA
*panagiotis.tsonis@notes.udayton.edu*



**Abstract** In this chapter, we have touched upon some of the key processes of vertebrate limb regeneration from the formation of the wound epithelium to pattern formation, to provide a picture of the many complex and intricate facets of this system. Our synthesis incorporates recent advances in molecular biology, which has revealed some important factors related to the initiation, induction and patterning in limb regeneration.

**Keywords** Vertebrates · Limbs · Regeneration

## 1 Overview

Regeneration of the adult limb is a fascinating phenomenon, which occurs only in the urodele amphibians. What makes these species so unique in their ability to regenerate their limbs? Why is this mechanism lost in higher vertebrates such as mammals? Understanding the intricate process of amphibian limb regeneration will hopefully shed some light on these questions. The process of limb regeneration requires a complex set of events leading to the formation of many different tissue types, such as bone, muscle, nerves, and blood vessels. At the end of this amazing process, an exact replica of the missing part is built. Although this chapter emphasizes the process of limb regeneration in urodele amphibians, a comparison will also be drawn with instances of limb regeneration in anuran amphibians and mammals.

## 2 Formation of the Wound Epithelium

Following amputation of the limb, the surface of the wound is covered rapidly with epithelial cells forming what is called the wound epithelium (WE). Within minutes amputation induces rapid signal transduction events [1]. The WE influences the underlying cells to dedifferentiate and re-enter the cell cycle. During this stage muscle, cartilage, and connective tissue lose their characteristics and become blastema cells. The blastema cells proliferate and form the blastema, a region of dedifferentiated mesenchymal cells, which ultimately give rise to the new limb (Fig. 1). The formation of the WE is a critical event, for without it regeneration does not take place [2]. The wound epithelium is distinct from that of the normal epithelium and is formed by the migration of epidermal cells at the edge of the amputation surface [3] (Fig. 2). Due to this distinction, it is believed that the expression of key molecules within the WE is paramount for regeneration. These key molecules may be important factors signaling the underlying tissues to re-enter the cell cycle and differentiate. One of these, WE3, is probably related to secretion [4]. We will now examine several molecules that are predominantly expressed within the wound epithelium.

Following the covering of the amputation surface, the cells of the WE begin to synthesize proteins essential for the basement membrane such as laminin and collagen type IV [5]. Collagen type XII has also been shown to be expressed within the wound epithelium. Expression of collagen type XII was first observed three days post-amputation in the basal layer of the wound epithelium. By day 10 both the basal layer of the WE and the mesenchymal cells were expressing collagen type XII. As regeneration progressed expression of collagen type XII continued to change until becoming restricted to the perichondrium in late digit stage [6].

Other important factors that may be necessary for the proper formation of the wound epithelium are the matrix metalloproteinases (MMPs). Matrix metalloproteinases play a pivotal role in matrix degradation, which is an important step in formation of the wound epithelium. It has been suggested that MMPs may be involved in the initial dedifferentiation of cells preparing for regeneration by breaking down the extracellular matrix [7]. One of them, the

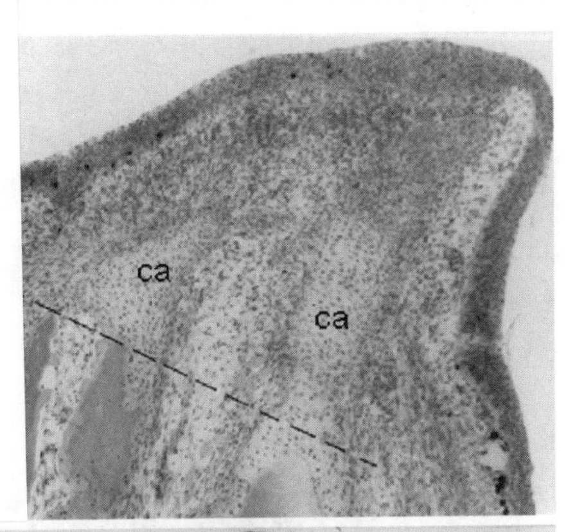


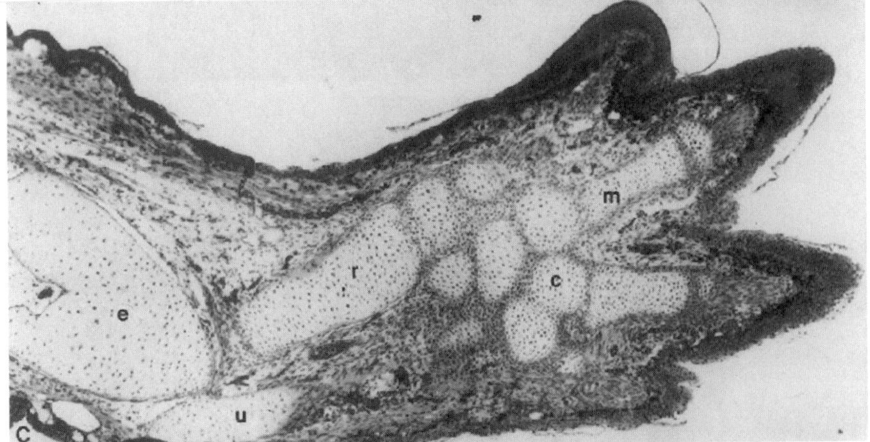


**Fig. 1A–C** Stages in the process of limb regeneration: **A** section through a blastema two weeks post-amputation, showing the accumulation of undifferentiated blastema cells – the *dotted line* denotes the plane of amputation; **B** section through a regenerating limb five weeks after amputation – note the differentiating cartilage (ca) composing the ulna and radius; **C** section through a regenerated limb two months after amputation at the elbow (e) level – note the regenerated skeletal elements and the correct formation of the four fingers – u: ulna, r: radius, c: carpals, m: metacarpals

newt MMPe, seems to be unique to newts and is specifically expressed in the apical epidermal cap and the wound epidermis [8]. Other MMPs, such as MMP3/10-b is expressed specifically at the basal layer of the apical ectodermal cap (AEC). MMP9 at a very early stage (2 h post-amputation) is localized in the wound epithelium and later is synthesized by cartilage and distributed in the basement membrane of the bone and the extracellular matrix of the blastema [8, 9]. During the early stages of limb regeneration within *Xenopus* (at early stages), gelatinase B (*Xmmp-9*) is expressed in the ectoderm and mesoderm at the tip of the amputated limb. This expression occurs very early during the process of limb regeneration, within 6–24 h after amputation [10]. It is impor-

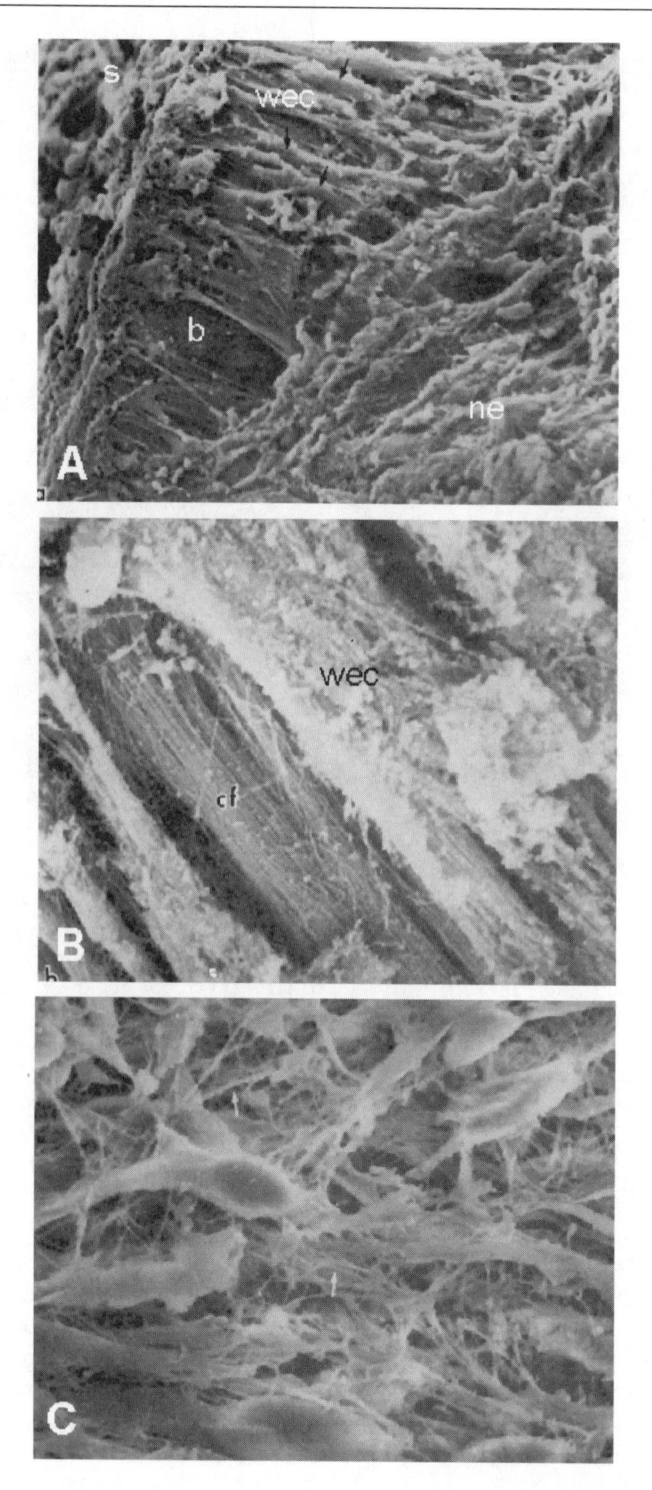
s
wec
b
ne
A
wec
cf
B
C

tant to note that *Xenopus* can regenerate their limbs only during early larval stages but lose this capacity after metamorphosis occurs.

Expression of complement factors shows an intriguing pattern. Complement component 3 (C3) was found present in the blastema and complement component 5 (C5) was exclusively found in the wound epithelium [11]. This might indicate that complement components might have non-immunologic functions in regenerative processes. Indeed, mice deficient in C5 show no liver regeneration [12].

We will see later (in relation to stimulation of limb regeneration) that growth factors, such as fibroblast growth factors (FGFs), have also been shown to be expressed within the wound epithelium. It is thought that these factors may provide the signals necessary for dedifferentiation and blastema formation.

## 3 Dedifferentiation

The formation of the WE appears to be what leads to the initiation of dedifferentiation and the formation of the blastema. Dedifferentiation and blastema formation are key events for successful limb regeneration. Fritsch [13] showed that the regenerate comes from the production of undifferentiated blastema cells (Fig. 2). Since that time, numerous studies have been performed examining the process of dedifferentiation during limb regeneration. The strongest evidence for dedifferentiation came from Hay and Fischman [14]. In their study, they observed the transition of muscle cells to mononucleated cells. These mononucleated cells contained large nuclei and ribonuclear granules both of which are indicative of protein synthesis. This knowledge proved that proteolysis, one of the dominating theories of the time, was not a contributing factor in the formation of blastema cells, since proteolysis does not involve active protein synthesis.

### 3.1 Transplantation Evidence

Transplantation of tissues or cells into X-ray irradiated limbs proved to be quite an instrumental supporter of the process of dedifferentiation. Irradiation has

**Fig. 2A–C** Scanning electron microscopy of wound closure and blastema cells: **A** dorsal view of an amputated limb showing the covering of the stump by the migrating cells (*arrows*) of the wound epithelium (wec), two days post-amputation – ne: normal epidermis, wec: wound epithelial cells, b: bone, s: stump: (×500); **B** higher magnification of A, showing the adhering WE cells (wec) on collagen fibers (cf) (×1800); **C** blastema cell population one week after amputation, indicating the fibroblast-like morphology of the blastema cells and the extensive extracellular matrix fibers (*arrows*)

been shown to inhibit regeneration. Dunis and Namenwirth [15] transplanted triploid cells into the limbs of irradiated diploid animals. The triploid cells participated in the regeneration process giving rise to cartilage, connective tissue, and fibroblasts. Steen [16] showed that triploid skin transplanted into irradiated diploid axolotls was able to form cartilage, connective tissue, dermis, and epidermis. In another set of transplantation experiments, Thornton [17] showed that metaplasia, transformation of one differentiated cell type into another, plays a role during regeneration. By removing the humerus and subsequently amputating through the area of the existing bone, limb regeneration took place with all skeletal elements distal to the humerus present. Regeneration took place despite the fact that the humerus was not present to be the source of the skeletal elements. These and other transplantation experiments help to solidify the importance of dedifferentiation in the formation of the blastema and ultimately the regenerate. For more details see [3].

## 3.2
## Cellular Evidence

Dedifferentiation has also been shown in vitro using cultured myotubes. Blastema explants in culture show both dedifferentiation and metaplasia. Polynucleated cells, from muscle explants, have been shown to go through dedifferentiation. When these cells were labeled and introduced into the blastema, they became mononucleated. In addition these polynucleated cells were also found to contribute to the regenerate, as the label was also present in cartilage, indicating metaplasia [18, 19]. Obviously these cells were able to re-enter the cell cycle. Retinoblastoma (Rb) has been found to be involved in such re-entry regulation and so is thrombin activation [20, 21]. To examine cell cycle re-entry and dedifferentiation/differentiation independently, cell cycle progression was inhibited through X-irradiation of the cells or transfection with p16 (CDK 4/6 inhibitor). It was found that post-mitotic newt myotubes generate mononucleated cells in the absence of cell cycle progression [22].

It has also been shown that mammalian post-mitotic nuclei re-enter the cell cycle. This was specifically shown when hybrid myotubes were created by the fusion of mouse C2C12 and newt A1 myogenic cells. Upon serum stimulation the C2C12 nuclei re-enter the cell cycle in the hybrids but not when grown alone [23]. Other work examining mammalian myotube dedifferentiation has shown that treatment of post-mitotic mammalian myotubes with an extract generated from newt regenerating limbs allowed for dedifferentiation. This indicates that when stimulated with the appropriate factors, mammalian myotubes can dedifferentiate, providing some light on the restriction of mammalian cells to regenerate [24]. Also, mammalian myotubes re-enter the cell cycle after transfection with the homeobox-containing gene *msx-1* [25].

# 4
# Stimulation of Regeneration

## 4.1
## Fibroblast Growth Factors

In addition to being expressed within the wound epithelium, fibroblast growth factors (FGFs) and their receptors (FGFRs) are thought to play a key role in providing important cues for proliferation and induction of limb regeneration. FGF-10 has been shown to be present in the mesenchymal cells and is correlated with regeneration in *Xenopus* [26]. Further investigation showed that introduction of FGF-10 protein into regeneration-deficient *Xenopus* limb buds was able to stimulate the expression of several genes including *shh, msx-1,* and *fgf-10.* In addition to gene expression, limb structures also formed at the regeneration-deficient stage following FGF treatment [27]. Along similar lines, previous research has shown that FGFR-1 and FGFR-2 are expressed in the blastema mesenchyme and wound epidermis of regeneration-competent *Xenopus* hindlimbs but not in the wound epidermis of regeneration-deficient hindlimbs. Inhibitors of these receptors suppress regeneration in regeneration-competent *Xenopus* hindlimbs [28].

FGF-2 has been proposed to be the elusive neurotrophic factor. It has been established that limb regeneration is dependent on an adequate nerve supply, suggesting that the nerves release a neurotrophic factor required for the proliferation of blastema cells [29]. In axolotl limb regeneration, it has been shown that, upon denervation, expression of FGF-2 and *Dlx-3* (a homolog of the homeobox-containing distal-less; see below) is down regulated in the epidermis. FGF-2 replacement, which is expressed in the nerves and the apical epidermal cap, re-establishes the normal expression of *Dlx-3* [30]. FGF-2 has also been shown to be an endogenous mitogenic factor responsible for blastema formation in frogs.

In *Xenopus* early larvae, however, hindlimb regeneration can proceed in the absence of nerves, while in late larvae this is not the case. It seems that following denervation, FGF-2 levels increase in regeneration-competent *Xenopus* limb buds but not in regeneration-deficient limb buds. This nerve-independent regeneration seen during the larval stages is due to the presence of FGF-2, which is serving as a neurotrophic-like factor. This suggests that the low level of FGF-2 present in the late *Xenopus* limb buds (regeneration deficient) is the reason for the onset of nerve dependence and ultimately why there is a lack of blastema formation at this stage [31].

Another implication of FGF2 as a factor in blastema formation came from a study in which the regeneration rates of two urodele species were examined. In this study, FGF-2 expression was examined in *T. carnifex* and *T.v. meridionalis.* Both species have the ability to regenerate their limbs, but the regeneration process is much faster in *T.v. meridionalis* than in *T. carnifex.* The rate of regeneration is correlated with delayed expression of FGF-2 in *T. carnifex.*

The delayed expression of FGF-2 resulted in the delayed formation of the apical ectodermal cap (AEC), which is necessary for blastema formation [32].

### 4.2 *Msx* Genes

Another set of genes being implicated in playing an essential role during limb regeneration is the *msx* genes. The *msx* genes are transcription factors expressed in regions of epithelial-mesenchymal interactions during embryogenesis of developing appendages. Numerous studies implicate these genes in the maintenance of embryonic tissues in an undifferentiated, proliferative state. Carlson et al. [33] examined the expression of *Msx-2* during limb development and limb regeneration in the axolotl. They found *Msx-2* expression to be downregulated at late stages of limb development, but re-expressed within one hour of limb amputation. These results indicate that *Msx-2* may have an important role in initiating the limb regeneration cascade. Earlier studies like the one described above have shown that, similar to *Msx-2, Msx-1* is also upregulated in response to limb amputation in the newt [34, 35].

Recent evidence has come about supporting the role of *msx* genes in mammalian digit regeneration. Mice as well as humans have the capacity to regenerate their digits upon amputation. Regeneration of the lost tip depends on the treatment given as well as the level of amputation. If the amputation is distal to the last interphalangeal joint and nothing is done to close the wound, regeneration will be successful [36, 37]. The area of digit regeneration seems to correspond to a region associated with the nail bed. Previous studies have shown that both *Msx-1* and *Msx-2* are expressed in association with the nail organ of neonatal digits as well as in developing digits [38]. During development it has been shown that *Msx-1* mediates a BMP signaling pathway that leads to the induction of the apical ectodermal ridge (AER). In a study by Han et al. [39], the role of *Msx-1* and BMP4 were examined. It was found that *Msx-1* mutant mice display a regeneration defect. Part of this defect included the downregulation of *BMP4* expression. Treatment with exogenous BMP4 was able to rescue the regeneration defect. These experiments provided the first functional evidence linking *Msx-1* and BMP signaling to the control of digit tip regeneration in mammalian fetal tissue [39]. As mentioned above the role of *msx-1* in dedifferentiation of mammalian myotubes strengthens its importance in regulating the process of dedifferentiation and blastema formation in urodeles [25].

## 5 Pattern Formation

So far, we have seen the importance of dedifferentiation and blastema formation in the process of limb regeneration. Now we will look at the process of pattern formation to gain an understanding of how an exact replica of the missing part is produced. During pattern formation, all three axes (anterior-posterior, proximal-distal, and dorsal-ventral) must be properly attained. Several theories have been suggested as to how pattern formation takes place and we will briefly mention them next.

### 5.1 Positional Information

Wolpert [40] proposed the idea of positional information stating that every cell possesses intrinsic information regarding its body position. During embryogenesis, cells acquire sets of coordinates that tell them where they are within the limb. Different ideas regarding the interpretation of these coordinates into positional information have been proposed and will be discussed briefly.

### 5.2 Boundary Model

This model proposed by Meinhardt [41] states that a diffusible morphogen produced at the boundary of two tissues can offer positional information. At the apical ectodermal ridge, the boundary between the anterior-posterior (AP) and the dorsal-ventral (DV) axes, the morphogen is produced and ultimately sets the proximal-distal (PD) axis.

### 5.3 Polarizing Zone Model

This model was proposed primarily due to the discovery of the zone of polarizing activity (ZPA) during chick limb development [42]. The ZPA is a region found in the posterior limb bud that is able to specify skeletal development ultimately playing a role in AP patterning. This model proposed by Tickle et al. states that positional information is acquired due to the amount of time spent in the polarizing zone [43].

Imokawa and Yoshizato suggested that the newt blastema contains a ZPA through their work on *Sonic hedgehog* (*shh*) [44]. In situ hybridization experiments indicated that *shh* was expressed within the mesenchymal cells of the posterior region of regenerating blastemas of newt limbs. Grafting experiments in which forelimbs were amputated at the mid-upper arm level supported this notion. At mid-bud stage the blastema was removed and grafted onto the original stump with a 180° rotation (reversal of the AP/DV axes). To reverse the AP axis, the blastema was grafted onto the contra-lateral stump without rotation. The grafted blastemas formed supernumerary limbs and expressed *shh* ectopically in a newly induced ZPA. These experiments suggest that AP patterning seen during limb regeneration is due to the regulation of *shh*-ZPA, similar to what is seen during limb development. Another study examining the role of *shh* during axolotl limb regeneration also found *shh* to have a role similar to that seen in development and that when *shh* was ectopically expressed using the vaccinia virus supernumerary digits as well as other hand and feet elements were formed [45].

## 5.4 Polar Coordinate Model

In this model, there is no diffusible factor required to obtain the positional information. Rather, coordinates regarding positional information are obtained through interactions with neighboring cells [46, 47]. According to this model each cell possesses positional information with respect to radius and circumference. Two cells which come in contact with one another during wound healing or grafting fill-in the missing positional values (both radial and circumferential) creating a properly patterned limb. This model made spectacular and accurate predictions on limb morphogenesis during regeneration (for details see [1]).

## 5.5 Retinoic Acid and Pattern Formation

Numerous studies have been performed examining the role of vitamin A derivatives such as retinol and all-*trans* retinoic acid on limb regeneration. Maden found that vitamin A derivatives, retinol palmitate or all-*trans*-retinoic acid, affected the proximal-distal axis [48]. Limbs amputated at the wrist, which would normally regenerate only the hand elements regenerated more proximal structures. He also found that the proximalizing effect of vitamin A was dose dependent. At a high concentration, vitamin A became inhibitory to limb regeneration. Similar proximalization results were also obtained by using an analogue of all-*trans*-retinoic acid, 9-*cis*-retinoic acid. In fact, 9-*cis*-retinoic acid had a more potent morphogenetic effect when compared to all-*trans*-retinoic acid [49].

Such work, based on the effects of retinoic acid, prompted speculation that positional information could be obtained depending on a chemical gradient, specifically a gradient of vitamin A along the proximal-distal axis, with a higher concentration in the proximal tissues. This gradient ultimately results in the proximalization effect of vitamin A. Treatment with vitamin A following amputation at the wrist would expose these tissues to a higher concentration of vitamin A. This increase in vitamin A would result in the formation of more proximal elements.

In fact, vitamin A has been found to alter chick limb bud patterning in a similar fashion to the ZPA. Recall from earlier, that the ZPA is a region found in the posterior limb bud that specifies skeletal development. Saunders and Gasseling discovered that transplantation of the ZPA into the anterior limb bud resulted in skeletal duplication along the AP axis [42]. The same result was seen after retinoic acid was inserted into the anterior margin of a developing chick limb bud [50].

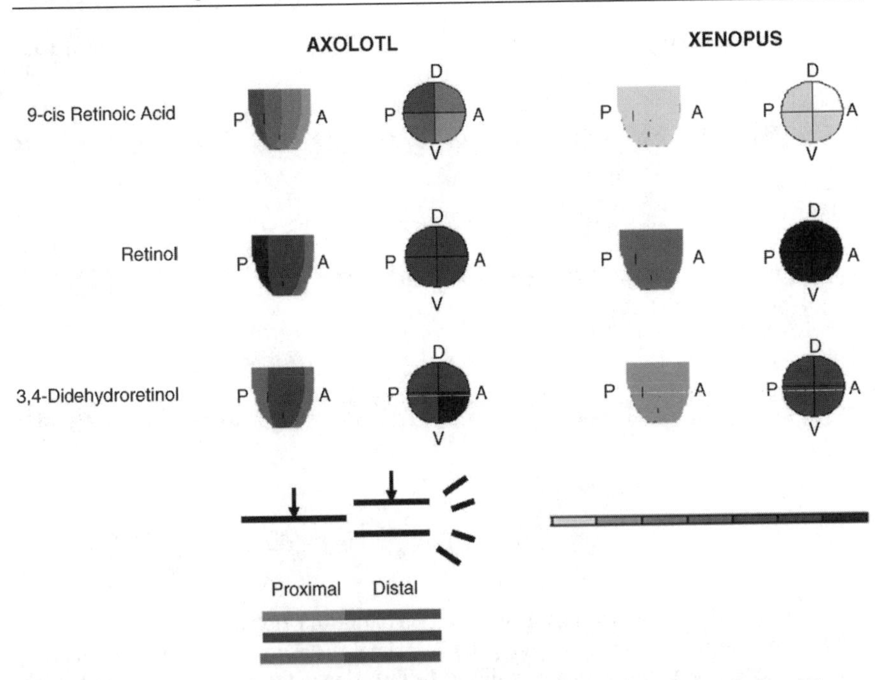


**Fig. 3** Illustrations showing the distribution of different retinoids in the axolotl and Xenopus blastema. The *different shades* indicate graded distribution ranging from low to high (see shade code). The shading is meant to show the qualitative difference and the presence of gradients and is not quantitative in regard of absolute values of retinoid concentrations. A: anterior, P: posterior, D: dorsal, V: ventral. The *lower left panel* shows the distribution of the retinoids along the proximal-distal axis

Of interest is the distribution of retinoids in the newt regenerating blastema. Scadding and Maden have shown that especially all-*trans* retinoic acid is present as a gradient along the anteroposterior axis with less at the anterior region and more at the posterior; as also seen in developing chick limb buds [51, 52]. Interestingly, there is more all-*trans* RA in the distal blastema as compared to the proximal, but this is valid only for the particular examined stage of regeneration. In Fig. 3 we present illustrations depicting the relative (qualitative) distribution of retinoids in newts and also compare with the absence of gradients in Xenopus limbs that are not able to regenerate. These studies suggest that a graded distribution of all-*trans* RA is correlated with the process of limb regeneration.

## 5.6 Hox Genes and Limb Regeneration

Hox genes are homeobox-containing genes, which control pattern formation and segmentation. The products of these genes are nuclear and contain a

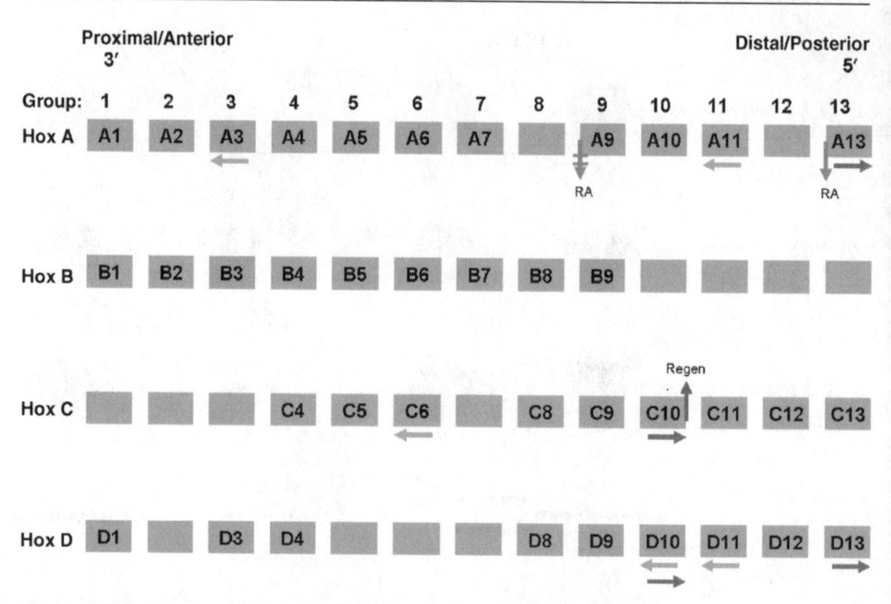


**Fig. 4** An illustration indicating the organization of the four Hox gene clusters. *Horizontal arrows* show graded expression along the proximal-distal axis or along the anterior-posterior axis. The *tips of the arrows* indicate the direction of the gradient from low to high. *Perpendicular arrows* indicate specific regeneration expression of C10 and regulation by retinoic acid

helix-turn-helix motif as their DNA binding domain, similar to that seen in bacterial repressors. This DNA binding domain is termed the homeodomain because it was found in many homeotic genes. The homeodomain is 60 amino acids in length and is highly conserved among different species. The Hox genes are arranged in clusters and are expressed in similar patterns in species ranging from Drosophila to mammals. Due to two duplications in vertebrates, there are four clusters of Hox genes, A, B, C, and D. The 3′ Hox genes mark the anterior/proximal parts of the body whereas the 5′ Hox genes mark the posterior/distal parts. Each cluster contains several genes denoted as A1, A2...A13 (this varies slightly depending on the cluster) (Fig. 4). During limb development the 5′ gene, Hox D13 is expressed most posteriorally and distally with Hox D12, -11, -10 appearing sequentially and with Hox D10 being the most anterior and proximal [54].

Several Hox genes have been shown to be specific to or upregulated within the blastema during limb regeneration [55–61]. In Fig. 4 we can see the specific regulation of some of these Hox genes. The horizontal arrows indicate regulation in the proximal versus distal regions (or anterior-posterior) of the limb and the tip of the arrow indicates higher levels of expression. The perpendicular arrows indicate specific regulation in the regenerating limb. This is the case of HoxC10 upregulation (long transcript), which otherwise is absent from the developing and intact limb [60]. Likewise, the arrow for A13 shows downregulation by retinoic acid, while A9 is unaffected [59]. Also, re-expression of A9 during regeneration is synchronous with A13, arguing against colinearity in the expression of Hox genes during regeneration, which is the case during limb development. Interestingly, some of the posterior/distal Hox genes (D10, D11, A11) show higher expression in the proximal blastemas. This might bear significance because during regeneration the pattern has to be respecified, however, from an already established adult limb and not an embryonic limb. In addition to the expression of Hox genes that are members of the clusters, other non-clustered Hox genes have been found to be expressed during limb regeneration. One of the best examples is the *msx* genes mentioned in an earlier section. Others include *Emx-2*, with a clear higher distal/lower proximal distribution in the epidermis and wound epithelium [58], and *distal-less* [56].

## 6 What the Future Will Bring

When magazines of popular science want to attract the attention of readers in articles of regeneration, their best bet is regeneration of the limbs. And rightly so, since limb regeneration is indeed a complex and fascinating phenomenon and known to occur in amphibia. But will we ever be able to regenerate amputated legs or arms? The most optimistic of us ask: If another vertebrate can regenerate, why are mammals unable to do the same? According to many observers all we need to do is to study the newt in detail, understand what induces and maintains the process of regeneration and apply this knowledge to mammals. This reductionistic approach, however, hides too many problems. There are fundamental issues that need to be addressed. First, regeneration must have a clear evolutionary cause. We know or understand next to nothing about it. Second, the physiology of advanced mammals is very different from that of the amphibians. Third, the urodeles are endowed with other remarkable regenerative capabilities, such as that of the eye, spinal cord, brain, or heart [62]. We cannot and must not isolate the limb field from all the others. It could very well be that what triggers regeneration in all the different organs in newts is a common signal. Likewise, we should not minimize the importance of the lowly salamanders when compared with the regenerative potency of the other strategy used in regeneration, which employs stem cells. Careful and systematic studies on all aspects of regeneration will lead to a better understanding and eventually to much anticipated regeneration therapies.

## References

1. Tsonis PA, English D, Mescher AL (1991) J Exp Zool 259:252
2. Thornton CS (1957) J Exp Zool 134:357
3. Tsonis PA (1996) Limb regeneration. Cambridge University Press, New York
4. Tassava RA, Johnson-Wint B, Gross J (1986) J Exp Zool 239:229

5. Del Rio-Tsonis K, Washabaugh CH, Tsonis PA (1992) Proc Natl Acad Sci USA 89:5502
6. Wei Y, Yang EV, Klatt KP, Tassava RA (1995) Dev Biol 168:503
7. Miyazaki K, Uchiyama K, Imokawa Y, Yoshizatio K (1996) Proc Natl Acad Sci USA 93:6819
8. Kato T, Miyazaki K, Shimizu-Nishikawa K, Koshiba K, Obara M, Mishima H, Yoshizato K (2003) Dev Dyn 226:366
9. Yang EW, Gardiner DM, Carlson MR, Nugas CA, Bryant SV (1999) Dev Dyn 216:2
10. Carinato ME, Walter BE, Henry JJ (2000) Dev Dyn 217:377
11. Kimura Y, Madhavan M, Call MK, Santiago W, Tsonis PA, Lambris JD, Del Rio-Tsonis K (2003) J Immunol 170:2331
12. Mastellos D, Papadimitriou JC, Franchini S, Tsonis PA, Lambris JD (2001) J Immunol 166:2479
13. Fritsch C (1911) Zool Jahrb Abt Physiol 30:377
14. Hay ED, Fischman DA (1961) Dev Biol 3:26
15. Dunis D, Namenwirth M (1977) Dev Biol 56:97
16. Steen TP (1973) Am Zool 13:1349
17. Thornton CS (1938) J Morphol 62:219
18. Lo DC, Allen F, Brockes JP (1993) Proc Natl Acad Sci USA 90:7230
19. Tsonis PA, Washabaugh CH, Del-Rio-Tsonis K (1995) Sem Cell Biol 6:127
20. Tanaka EM, Gann AA, Gates PB, Brockes JP (1997) J Cell Biol 136:155
21. Tanaka E, Drechsel DN, Brockes J (1999) Curr Biol 9:782
22. Velloso CP, Kumar A, Tanaka EM, Brockes JP (2000) Differentiation 66:239
23. Velloso CP, Simon A, Brockes JP (2001) Curr Biol 11:855
24. McGann CJ, Odelberg SJ, Keating MT (2001) Proc Natl Acad Sci USA 98:13699
25. Odelberg SJ, Kollhoff A, Keating MT (2000) Cell 103:1099
26. Yokoyama H, Yonei-Tamura S, Endo T, Izpisua Belmonte JC, Tamura K, Ide H (2000) Dev Biol 219:18
27. Yokoyama H, Ide H, Tamura K (2001) Dev Biol 233:72
28. D'Jamoos CA, McMahon G, Tsonis PA (1998) Wound Repair Regen 6:388
29. Singer M, Mutterperl E (1963) Dev Biol 7:180
30. Mullen LM, Bryant SV, Torok MA, Blumberg B, Gardiner DM (1996) Development 122:3487
31. Cannata SM, Bagni C, Bernardini S, Christen B, Filoni S (2001) Dev Biol 231:436
32. Giampaoli S, Bucci S, Ragghianti M, Mancino G, Zhang F, Ferretti P (2003) Proc R Soc Lond B 270:2197
33. Carlson MRJ, Bryant SV, Gardiner DM (1998) J. Exp Zool 282:715
34. Crews L, Gates PB, Brown R, Joliot A, Foley C, Brockes JP, Gann A (1995) Proc R Soc Lond 259:161
35. Simon HG, Nelson C, Goff D, Laufer E, Morgan BA, Tabin C (1995) Dev Dyn 202:1
36. Illingworth CM (1974) J Pediatr Surg 9:853
37. Borgens RB (1982) Science 217:747
38. Reginelli AD, Wang YQ, Sassoon D, Muneoka K (1995) Development 121:1065
39. Han M, Yang X, Farrington JE, Muneoka K (2003) Development 130:5123
40. Wolpert L (1969) J Theor Biol 25:1
41. Meinhardt H (1983) J Embryol Exp Morphol 76:139
42. Saunders JW, Gasseling M (1968) In: Fleischmayer R, Billingham RE (eds) Epithelial-mesenchymal interaction. Williams and Wilkins, Baltimore. pp78
43. Tickle C, Summerbell D, Wolpert L (1975) Nature 254:199
44. Imokawa Y, Yoshizato K (1997) Proc Natl Acad Sci USA 94:9159
45. Roy S, Gardiner DM, Bryant SV (2000) Dev Biol 218:199
46. Bryant PJ, Bryant SV, French V (1977) Sci Am 237:66

47. Bryant SV, French V, Bryant PJ (1981) Science 212:993
48. Maden M (1982) Nature 295:672
49. Tsonis PA, Wahabaugh CH, Del Rio-Tsonis K (1994) Roux's Arch Dev Biol 203:230
50. Tickle C, Alberts BM, Wolpert L, Lee J (1982) Nature 296:56
51. Scadding SR, Maden M (1994) Dev Biol 162:608
52. Thaller C, Eichele G (1987) Nature 327:625
53. Removed
54. Izpizua-Belmonte JC, Tickle C, Dolle P, Wolpert L, Duboule D (1991) Nature 350:585
55. Brown R, Brockes JP (1991) Development 111:489
56. Beauchemin M, Savard P (1992) Dev Biol 154:55
57. Beauchemin M, Noiseux N, Tremblay M, Savard P (1994) Int J Dev Biol 38:641
58. Beauchemin M, Del Rio-Tsonis K, Tsonis PA, Trembley M, Savard P (1998) J Mol Biol 279:501
59. Gardiner DM, Blumberg B, Komine Y, Bryant SV (1995) Development 121:1731
60. Carlson MR, Komine Y, Bryant SV, Gardiner DM (2001) Dev Biol 229:396
61. Christen B, Beck CW, Lombardo A, Slack JMW (2003) Dev Dyn 226:349
62. Tsonis PA (2002) Differentiation 70:397

Received: January 2004

Adv Biochem Engin/Biotechnol (2005) 93: 83–100
DOI 10.1007/b99972


# Mammalian Fetal Organ Regeneration

Amy S. Colwell · Michael T. Longaker · H. Peter Lorenz (✉)

Department of Surgery, Pediatric Surgical Research Laboratory, Stanford University School of Medicine, 257 Campus Drive, Stanford CA94305-5148, USA
*plorenz@stanford.edu*

**Abstract** The developing fetus has the remarkable ability to heal dermal skin wounds by regenerating normal epidermis and dermis with restoration of the extracellular matrix architecture, strength, and function. The biology responsible for scarless wound healing in skin is a paradigm for ideal tissue repair. This regenerative capacity is lost in late gestation when fetal wounds heal with fibrosis and scar. Early in gestation, fetal skin is developing at a rapid pace in a unique environment. Investigation of normal skin embryogenesis and comparison between early scarless and late scarring fetal wounds has revealed distinct differences in inflammatory response, cellular mediators, wound contraction, cytokines, growth factors, and extracellular matrix modulators. The knowledge gained from comparative observational studies has served as a base for experimental interventions in animal models to induce or ameliorate scar. Although much progress has been made over the past decade, the mechanism of fetal wound healing remains largely unknown and attempts to mimic the scarless wound phenotype have not been completely successful. Identification of more key genes involved in skin regeneration may have implications in adult skin wounds and repair in other organ systems.

**Keywords** Fetal wound healing · Scarless repair

**List of Abbreviations**

| | |
|---|---|
| DDR | Discoidin domain receptor |
| ECM | Extracellular matrix |
| FGF | Fibroblast growth factor |
| HA | Hyaluronic acid |
| HIF | Hypoxia-inducible factor |
| HSP | Heat shock protein |
| IL | Interleukin |
| MAPK | Mitogen-activated protein kinase |
| MMP | Matrix metalloproteinase |
| PAI-1 | Plasminogen activator inhibitor 1 |
| PCR | Polymerase chain reaction |
| PDGF | Platelet derived growth factor |
| RTK | Receptor tyrosine kinase |
| TGF-beta | Transforming growth factor – beta |
| TIMP | Tissue inhibitor of metalloproteinase |
| uPA | Urokinase plasminogen activator |
| VEGF | Vascular endothelial growth factor |

## 1 Fetal Skin Anatomy, Development, and Wound Healing

The skin is the mammal's largest organ and outermost covering. It primarily functions as a barrier to prevent desiccation, toxin entry, and microbial infection [1].

Fetal skin structure and histology change dramatically with development. In early gestation, mutual inductive mechanisms between ectoderm and mesoderm stimulate development of the epidermis and dermis. Human epidermal primordial cells derived from ectoderm proliferate at seven weeks gestation forming a squamous layer of periderm and a basal germinative layer. The periderm cells are keratinized, shed, and eventually replaced by the stratum corneum at 21 weeks. The basal germinative layer becomes the stratum germinativum, a source of new cells for dermal appendages, the intermediate layers found in mature skin, and hair germs. Hair follicles begin development in the 9th–12th week. Peripheral follicular cells become the epithelial root sheath and surrounding mesenchymal cells form the dermal root sheath. Mesoderm-derived mesenchymal cells produce collagen and elastic connective tissue fibers of the dermis beginning at 11 weeks. Mesenchyme also differentiates to form arteries and veins around tubular endothelial cell structures. The muscular coat of arteries and veins may arise from myofibroblasts. Further skin maturation with dermal extracellular matrix thickening continues into the postnatal period [2].

Mature skin consists of a thin epidermis, which is nourished and maintained by the underlying thicker dermis. The epidermis and hair follicles have prolific growth characteristics exemplified by their continual renewal and turnover [1]. In the epidermis, epithelial stem cells reside in the basal layer, whose progeny begin the process of terminal differentiation and migrate upward to form the stratified layers. For hair shaft renewal, progeny of epithelial stem cells residing in the hair bulb migrate to the base of the follicle and surround the dermal papilla in the matrix. Signals from the papilla, in turn, induce these daughter cells to begin the differentiation process, forming the hair shaft and inner root sheath [3].

The mammalian fetus is unique in that injury during skin development early in gestation heals in a scarless pattern with fine reticular collagen and regrowth of hair follicles. This cutaneous regenerative capacity, however, is not retained in the postnatal and adult mammal. Late fetal and postnatal injury causing disruption of the collagen architecture results in regrowth of disorganized collagen bundles and no hair follicle regrowth, a process termed scar formation.

## 2
## Experimental Methods and Models of Scarless Wound Repair

Tissue and organ regeneration research in the mammalian fetus has focused primarily on the integumentary system. Skin wounds are relatively easy to create and monitor in vivo. Injury models include incisional (closed) and excisional (open) wounds. Different techniques and wounding patterns can help elucidate answers to specific questions. For example, an incisional fetal lamb wound model demonstrated that the fetal environment could not induce scarless repair in adult skin. Adult sheep skin was transplanted onto the backs of 60-day fetal lambs (term 145 days) where they were perfused by fetal blood and bathed with amniotic fluid. Incisional wounds in the adult skin grafts healed with scar while incisional wounds in the nearby fetal skin healed scarlessly [4].

Many different animal models have been developed to investigate fetal skin regeneration and include the following large and small animals: mouse, rat, rabbit, opossum, lamb, and monkey. The wounded tissue specimens can be evaluated by gross examination, histological sectioning, immunohistochemistry, confocal microscopy, immunoblotting, polymerase-chain reaction (PCR), and microarray technology. Additionally, individual cells or organs in the wound environment can be grown in cell or organ culture and manipulated in vitro.

### 2.1 Large Animal Models

Large animal models such as the lamb have a long gestation (>145 days) and thus a longer intrauterine post-operative period. Because of their size, fetal surgical manipulation is technically simpler. However, these animals are much more expensive to obtain and house. Additionally, fewer molecular tools can be developed for experimentation because their genomes are less characterized.

### 2.2 Small Animal Models

Small animals such as the rat, mouse, and rabbit are relatively inexpensive and have short gestational periods. Murine models enjoy the advantages of a well-characterized genome that is amenable to molecular investigation and creation of knockout and transgenic animals [5]. Drawbacks to the small animal models include a short interval where fetal manipulation is feasible and also a short intrauterine post-operative period. Their size makes surgical manipulation more technically challenging.

## 3 Scarless Wounds: Tissue Specificity

The remarkable ability of fetal skin wounds to heal scarlessly has naturally inspired broader interest in the applicability of this phenomenon to other organ systems. In short, this regenerative capacity seems to be limited to fetal skin, although some interesting preliminary observations have been made in incisional wounds in organ culture systems. Most fetal in vivo injury experiments in non-skin tissues have shown scar formation.

### 3.1 Fetal Stomach, Intestine, and Diaphragm Wounds Heal with Scar

Full thickness incisional gastric and diaphragm wounds in fetal lambs heal with pronounced scar formation and dense intraperitoneal fibrous adhesions [6]. Likewise, laparotomy and intestinal enterotomy in fetal rabbits show a similar degree of fibrosis and neovascularization when compared to adult rabbits [7, 8].

### 3.2 Stabilized Fetal Long Bone Fractures Heal More Rapidly and with Less Callus than Adult Fractures

Animal models of fetal fractures in rabbits and chicks show rapid fracture healing with a variable amount of callus formation [9]. When motion is controlled by rigid internal fixation, either by natural stabilization of ulnar fractures by an intact radius or with miniplates, primary bone healing with minimal callus occurs [9, 10].

### 3.3 Incisional Wounds in Fetal Murine Heart and Lung Organ Culture Heal with Regeneration

Inspired by the fetal integument wound healing models, Blewett et al. devised murine lung and heart explant culture systems to examine wound repair [11, 12]. Linear incisions in 14- to 18-day fetal lungs examined 7 days after wounding showed restitution of tissue architecture. Collagen scar deposition was undetectable by trichrome staining. Similarly, incisional wounds in 14-day murine fetal hearts in organ culture heal by regeneration and cardiomyocyte migration. At 18 days, a transitional wound phenotype occurs with incomplete bridging of the wound by cardiomyocytes and no collagen deposition. By 22 days, the hearts demonstrate repair of wounds by scarring with granulation tissue and collagen accumulation. Although promising, in vivo models subject to mechanical stresses and an intact blood supply are needed to confirm these results.

## 4 Skin Regeneration: Scarless Fetal Wound Phenotype

### 4.1 Structural Proteins – Collagen

Collagen is the dominant structural protein in all human connective tissue. Although several types of collagen exist, Type I predominates and is the principal component of both adult and fetal ECM [13]. Its strength is derived from a triple helix configuration of polypeptide chains, which are cross-linked and stabilized by lysyl oxidase. Fetal skin has a higher ratio of Type III to Type I collagen than adult skin[14]. This ratio gradually decreases with maturation and resembles the adult phenotype in the postnatal period [15]. In scarless fetal wounds, collagen is rapidly deposited in a fine reticular pattern indistinguishable from uninjured skin. In contrast, adult scarring wounds have disorganized Type I collagen bundles with more collagen cross-linking [16–18] (Figs. 1–4). Analysis of adult rat wounds with reverse transcriptase-polymerase chain

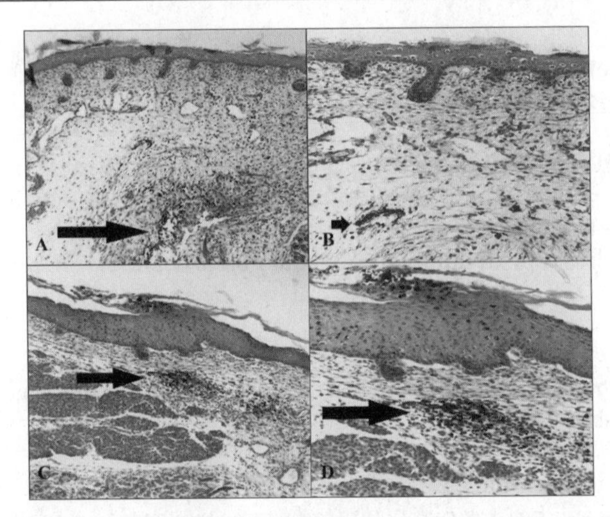


**Fig. 1A–D** Scarless healing of E16 fetal wounds (H&E stain). *Black arrows* indicate India ink tattoo made at the time of wounding in order to demonstrate scarless wound location: **A, C** healed wounds at 72 h (×100). The epidermal appendage (developing hair follicles) pattern shows numerous appendages directly in the healed wound; **B, D** magnified views of the same wounds showing epidermal appendages within the wound site (×200). No inflammatory infiltrate is present

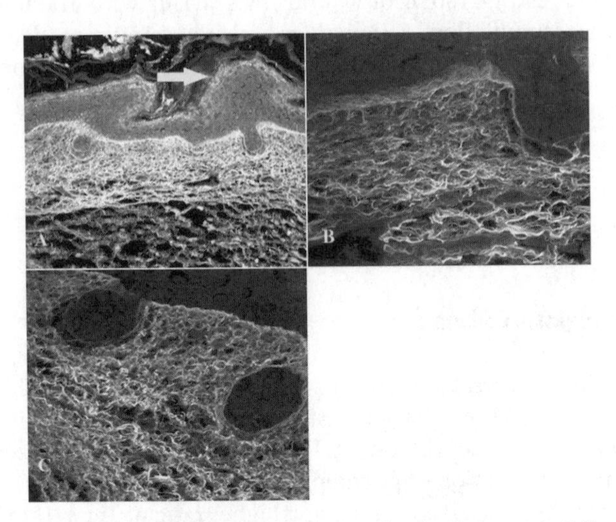


**Fig. 2A–C** Scarless healing of E16 fetal wounds (confocal microscopy). Collagen fibers are stained with sirius red and appear white: **A** healed wound harvested at 72 h (×200). The epidermis is thickened at the wound site (*arrow*). The collagen fiber is reticular and unchanged from the surrounding dermis; **B** healed wound harvested at 72 h under a higher magnification (×1000). The collagen fibers are thin and closely approximating each other with little interfiber space. The fibers are arranged in a wispy reticular pattern; **C** nonwounded E19 skin at the same magnification as B (×1000). The dermal collagen fiber pattern is identical to B

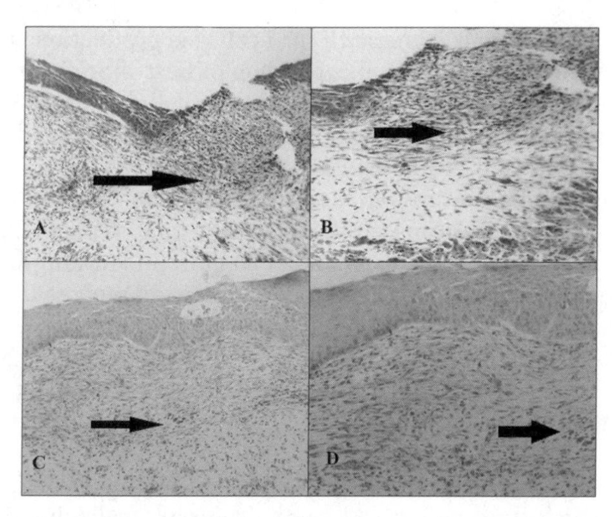


**Fig. 3A–D** Scar formation after transition point in E18 fetal wounds (H&E stain). *Black arrows* indicate green vital dye tatoo made at the time of wounding: **A** wound at 24 h. The wound remains open and an inflammatory cell infiltrate is present (×100); **B** magnified view (×200) of the incompletely healed wound at 24 h as shown in A; **C** wound at 72 h. No epidermal appendages are present, consistent with adult-type repair and scar formation (×100); **D** magnified view (×200) of wound shown in C, demonstrating the lack of dermal appendages and an inflammatory cell infiltrate

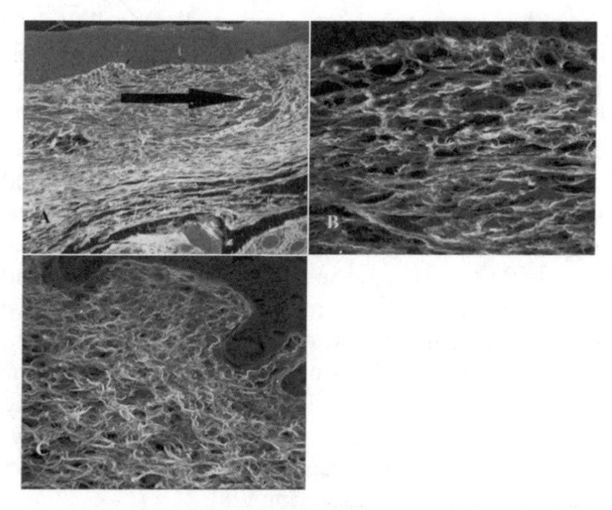


**Fig. 4A–C** Scar formation after transition point in E18 wounds (confocal microscopy). Collagen fibers are stained with sirius red and appear white: **A** healed wound at 72 h (×200). The wound dermal collagen pattern is different from the surrounding non-wounded dermis. The fibers are less densely compacted. No epidermal appendages are present. Neo-vascularization is shown with the *black arrow*; **B** healed wound at 72 h at a higher magnification (×1000). The collagen fibers are thicker, but with greater interfiber spaces compared to non-wounded dermis; **C** non-wounded skin at E21 days gestational age (×1000). When compared to wound collagen fibers (B), non-wounded dermal collagen fibers are thinner with less interfiber space

reaction (RT-PCR) reveals increased Type I collagen production after wounding. This correlates with increased expression of heat shock protein 47 (HSP 47), a molecular chaperone specifically involved in collagen processing. In contrast, fetal wounds showed no difference in collagen I production or HSP 47 expression [19]. Lovvorn et al. implanted PVA sponges in fetal sheep wounds and noted increased collagen cross-linking with advancing gestational age that paralleled the transition from scarless to scar-forming repair [20]. Although Type I collagen cross-linking is essential for adult wound healing and strength, its rigidity may impede the movement of cellular mediators required for rapid cellular regeneration in the fetus.

## 4.2 Proteoglycans and Adhesion Proteins

Hyaluronic acid (HA) is a negatively charged, unsulfated glycosaminoglycan of the ECM which traps and impedes water molecules allowing resistance to deformation and facilitating cellular movement [13]. The hyaluronic acid (HA) content of scarless fetal wounds increases more rapidly, is more sustained, and is overall greater than that of adult wounds [21]. Fetal wounds have greater HA stimulating activity and fewer pro-inflammatory cytokines, such as interleukin 1 (IL-1) and tumor necrosis factor-alpha, that down-regulate HA expression [22].

Scarless fetal wounds are characterized by a more rapid upregulation of ECM adhesion proteins and differential expression of cell surface receptors (integrins). The adhesion protein fibronectin mediates cellular attachment to the ECM and attracts fibroblasts, keratinocytes, and endothelial cells to the site of injury [23]. Early gestation fetal rabbit wounds express fibronectin 4 h after wounding while expression is not seen until 12 h in the adult [24]. Whitby et al. found no difference in the onset of expression, but noted more sustained expression in the adult [17]. Tenascin blocks fibronectin-mediated cellular attachment. In upper lip wounds of mice, tenascin appears more rapidly in the fetus (1 h) compared to the adult (24 h) and precedes cellular migration [17]. This suggests a role for tenascin in the rapid closure of fetal wounds. Collagen integrin receptors in fetal fibroblasts are differentially expressed with increasing gestational age. Integrins are involved in cellular attachment to the ECM and intracellular signaling. Fetal fibroblasts have increased alpha 2 integrin subunit expression and decreased alpha 1 and 3 integrin subunit expression compared to the adult. This correlates with the low capacity of fetal fibroblasts to contract a collagen gel and may have implications in the differences seen between fetal and adult wound contraction [25].

## 4.3 Extracellular Matrix Modulators

Regulators of collagen organization and degradation influence the ECM architecture. Decorin, a modulator of collagen fibrillogenesis, shows no change in fetal wounds but is up-regulated in adult wounds [26]. Fibromodulin-facilitated cellular migration is down-regulated in adult wounds and unchanged in the fetal wound. [27]. This may prove useful as a marker of wound phenotype – if exogenous factors decrease scarring, they may decrease decorin and increase fibromodulin.

Proteases such as matrix metalloproteinases (MMPs) and tissue-derived inhibitors (TIMPs) function in ECM turnover. Levels of MMP1 and MMP9 are greater and increase more rapidly in scarless compared to scarring wounds. MMP14 expression increases threefold in scarless wounds but is unchanged in scarring wounds. In contrast, scarring wounds have up-regulation of TIMP-1 and TIMP-3 expression. Overall, scarless wounds have a higher ratio of MMP to TIMP expression, suggesting rapid degradation of collagen is important to prevent over-accumulation and scar [28].

Plasmin is another protease, derived from plasminogen by urokinase-type plasminogen activator (uPA), which plays a central role in wound repair and fibrosis. The activation of plasminogen to plasmin by uPA is inhibited by plasminogen activator inhibitor-1 (PAI-1), an enzyme associated with postnatal organ fibrosis. Scarless fetal murine wounds have a higher ratio of uPA to PAI-1 compared to scarring wounds [29]. The addition of aprotinin, a serine protease inhibitor, causes an increase of collagen deposition in the scarless fetal wounds. The higher ratio of protease to inhibitors in scarless fetal wounds likely favors higher ECM turnover, which facilitates migration of fetal cells and promotes scarless repair.

## 4.4 Transition Wound: The Earliest Scar

There is likely a developmentally regulated threshold for scarless healing based on gestational age and the extent of injury. The ontogenetic transition of rat skin has been defined in an organ culture system and confirmed in vivo with confocal microscopic analysis [18, 30]. This transition point lies between days 16.5 and 18.5 of gestation (term=21.5 days). In a human fetal skin model, the transition point occurs after 24 weeks of gestation [31].

Wound size modulates the transition point. In fetal lambs, increasing wound size increased the frequency of scarring at a gestational age when smaller wounds healed scarlessly [32]. In nonhuman primates, the transition from scarless to scarring repair has been shown to proceed through an intermediate wound phenotype. Fetal monkey lip incisional wounds heal with restoration of normal epidermal appendage and dermal collagen architecture in midgestation. Hair follicle regeneration is an integral component of scarless repair. At the start of the third trimester, these wounds do not restore epidermal appendage (hair follicle and sebaceous gland) architecture, but still heal with a normal collagen dermal pattern. Thus, a "transition wound" phenotype occurs. By the mid-third trimester, the wounds heal with a typical scar pattern – no appendages and collagen scar [33].

# 5 Mechanisms of Scarless Repair

## 5.1 Fetal Wounds have Decreased Inflammatory Cells

Fetal wounds heal rapidly with a paucity of inflammatory cells. This key observation has stimulated interest in the role of cellular inflammatory mediators, cytokines, and growth factors in fetal wound healing. The absence of an acute inflammatory infiltrate in scarless wounds may be partly explained by decreased fetal platelet degranulation and aggregation. Although there is no difference in size, organization, or granule content by transmission electron microscopy in fetal compared to adult platelets, fetal platelets have less platelet-derived growth factor (PDGF), TGF-beta1, and TGF-beta 2 than their adult counterparts [34]. Fetal platelet exposure to collagen in vitro does stimulate growth factor release; however, the platelets still do not aggregate [35]. Olutoye et al. further investigated the aggregatory capabilities of adult and fetal porcine platelets after exposure to collagen and adenosine diphosphate. The fetal platelets responded suboptimally to collagen and showed an age-dependent aggregatory response to ADP exposure corresponding with the transition period for cutaneous scarless to scar-forming wounds [36]. HA also suppresses aggregation and release of PDGF from fetal platelets in a dose-dependent fashion, having the greatest effect in the HA-rich fetal environment [37]. Additionally, fewer neutrophils are present in the fetal wound, and an age dependent defect in the ability of fetal neutrophils to phagocytose pathogenic bacteria has been demonstrated in fetal sheep [38].

## 5.2 Differential Expression of Cytokines Modulates Scar Formation

### 5.2.1 Transforming Growth Factor – beta (TGF-beta)

The transforming growth factor-betas were linked to wound healing shortly after their discovery more than twenty years ago. TGF-beta is chemotactic for fibroblasts, keratinocytes, and inflammatory cells. It also stimulates collagen I production by fibroblasts [39]. TGF-beta expression is increased by wounding and injury. Expression is modified by decorin, fibromodulin, hypoxia, and hypoxia-inducible factor-1 alpha (HIF-1alpha). Decorin and fibromodulin bind TGF-beta isoforms and may regulate their bioactivity by sequestering TGF-beta ligand in the extracellular matrix [40]. TGF-beta 3 expression has been shown to be hypoxia-inducible in mouse embryonic fibroblasts. However, TGF-beta three levels are nearly undetectable in HIF-1 null fibroblasts (HIF-1–/–), suggesting HIF may be an upstream regulator of TGF-beta three expression [41]. In turn, TGF-beta decreases MMP expression, which favors collagen matrix accumulation after injury [39].

Evidence implicating TGF-beta 1 as a pro-scarring cytokine is well established. Immunohistochemical analysis reveals no change in TGF-beta 1 and 2 expression in fetal rabbit wounds but increased expression in adult wounds [42]. Scarless wounds in fetal mice have less TGF-beta 1 staining than neonatal or adult wounds [43]. Furthermore, the relative proportion of TGF-beta isoforms, and not the absolute amount of any one isoform, may determine the wound phenotype. In scarless fetal rat wounds, TGF-beta 1 and TGF-beta 2 expression is decreased while TGF-beta 3 expression is prolonged. Conversely, TGF-beta 1 and TGF-beta 2 expression is increased while TGF-beta 3 expression is decreased and delayed in scarring fetal wounds [44]. Expression of TGF-beta receptor type I and type II is also increased in scarring fetal rat wounds compared with scarless wounds. Thus, increased TGF-beta 1 and TGF-beta 2 accompanied by decreased TGF-beta 3 is associated with repair that results in scar [45].

TGF-beta 3 may have additional roles in development of the palate as evidenced by 100% clefting in TFG-beta 3 knock-out homozygous pups. This may be useful as an intrauterine model of cleft palate; however, early perinatal mortality prohibits postnatal applications [46].

### 5.2.2 Other Growth Factors and Interleukins

Platelet-derived growth factor (PDGF) and fibroblast growth factor (FGF) are additional pro-fibrotic cytokines. PDGF, a potent mitogen and chemoattractant for fibroblasts, has prolonged expression during scar formation but has little expression in fetal wounds [47]. The FGF family of cytokines and receptors, including keratinocyte growth factors 1 and 2, increases expression in fetal skin during development and in adult skin during wound repair [48]. Dang et al. found decreased expression of FGF-1 and FGF-2 in scarless rat fetal wounds [48]. In addition, FGF receptor 2 had earlier, more sustained down-regulation when compared to scarring fetal wounds. Thus, many cytokines associated with scar formation in adult wounds are decreased during scarless repair.

In contrast, a mitogen for endothelial cells, vascular endothelial growth factor (VEGF), increases two-fold in scarless wounds while its expression remains unchanged in scarring fetal wounds [49]. An increased stimulus for angiogenesis and vascular permeability may assist in the rapid healing of fetal wounds.

Interleukins are cytokines important in chemotaxis and activation of inflammatory cell mediators. IL-6 stimulates monocyte chemotaxis and macrophage activation while IL-8 attracts neutrophils and stimulates neovascularization [50]. In scarless fetal wounds, little or no inflammation occurs. In adults, wounding stimulates a rapid and prolonged increase in IL-6 and IL-8 [50, 51]. However, only a short and limited expression is induced in fetal wounds. Both IL-6 and IL-8 expression are significantly lower in early gestation fetal fibroblasts at baseline and with PDGF stimulation compared to in adult fibroblasts

[50]. Thus, the fetal fibroblast likely expresses less pro-inflammatory mediators after injury compared to the adult.

## 5.3 Fetal and Adult Fibroblasts are Inherently Different

Synthesis and remodeling of the ECM by fibroblasts is essential for wound healing. Adult and fetal fibroblasts are recruited to the site of injury by soluble chemoattractants released by macrophages and neutrophils [13]. Fetal wounds characteristically have less inflammatory cells and cytokine expression yet heal more rapidly than adult wounds. This may be partly explained by intrinsic differences between adult and fetal fibroblasts.

Fetal and adult fibroblasts display differences in their production of collagen, HA, and other ECM components. In vitro, fetal fibroblasts synthesize more Type III and IV collagen than their adult counterparts, correlating with an increase in prolyl hydroxylase activity, the rate-limiting step in collagen synthesis [52]. Collagen synthesis is delayed in the adult wound while fibroblasts and progenitor cells infiltrate and proliferate. In contrast, fetal fibroblasts simultaneously proliferate and synthesize collagen [13]. Increases in cell density diminish HA production in the adult but has no effect on fetal fibroblast HA synthesis [53]. The fetal fibroblast is programmed to synthesize dermal ECM continuously during development and appears to have little change after injury.

Fetal fibroblasts likely have a greater ability to migrate into wounds, evidenced by their greater migration into collagen gels compared with adult fibroblasts. A migration stimulation factor secreted by fetal fibroblasts is purported to be responsible for this enhanced migratory ability [54]. Fetal fibroblasts have more surface receptors for hyaluronic acid, which also serves to enhance fibroblast migration [53]. Additionally, TGF-beta 1, which inhibits migration of confluent fibroblasts in vitro, is decreased in the fetal wound [55].

Differences in contractile fibroblasts, termed "myofibroblasts", have also been reported. Myofibroblasts, detected by the presence of alpha smooth muscle actin, appear in the adult wound one week after wounding. The adult wound content of myofibroblasts is greatest during the second to third week after injury and then decreases with time [13]. Fetal wounds made early in gestation have virtually no myofibroblasts. In contrast, scarring fetal and postnatal wounds have progressively more active myofibroblasts, which correlates with contraction and degree of scarring [56].

## 5.4 Differential Signal Transduction Pathway and Transcription Factor Expression

Efforts toward defining the scarless fibroblast phenotype have examined cellular signaling via receptor tyrosine kinase phosphorylation patterns and adapter protein, Shc, expression. Shc couples receptor tyrosine kinase to mitogen-activated protein kinase (MAPK) [57]. It serves as a key intermediate for discoidin domain receptor (DDR) signaling and may contribute to hypoxia-induced HIF protein stabilization and endothelial migration [58, 59]. Although TGF-beta signaling is predominantly mediated through serine/threonine kinase receptors, tyrosine kinase receptor signaling controls fundamental reaction sequences leading to gene activation [60]. Different receptor tyrosine kinase (RTK) phosphorylation patterns are observed between fetal and adult rat fibroblasts with increased amounts of epidermal growth factor receptor, DDR, and Shc proteins in fetal fibroblasts suggesting that RTK signaling may play a role in scarless repair [60].

Ultimately, the mechanistic differences between scarless and scarring repair may be regulated at the gene expression level. Homeobox genes are transcription factors that are implicated in the patterning and cell type specification events during development. These genes determine the direction taken by major developmental pathways, which involve activity of hundreds of genes. Their role in skin embryogenesis and wound healing is being investigated. Human homeobox genes MSX-1, MSX-2, and MOX-1 are differentially expressed in skin development [61]. MSX-1 and MSX-2 are detected in the epidermis, dermis and dermal appendages of developing fetal skin. In contrast, their expression is limited to epithelially-derived structures in the adult. MOX-1 expression is similar to MSX in early gestation while it is completely absent in adult cutaneous tissue. Additionally, human fetal scarless repair is associated with decreased expression of HOXB13 and increased PRX-2 expression [62]. Given that scarless repair is inherent to developing skin, it seems likely that coordinated control of groups of genes by transcription factors, such as Homeobox genes, has a crucial function during the repair process.

## 6 Experimental Interventions to Induce or Ameliorate Scar

The scarless fetal wound can be used to test whether cytokines, or other agents, have profibrotic or antifibrotic function. Using the knowledge of the scarless wound phenotype and mechanistic differences between fetal and adult wound healing patterns, investigators have manipulated both the scarless and scarring wound in attempts to elucidate the mechanisms of scarless repair.

### 6.1 Addition of Growth Factors or Inflammatory Stimuli Induces Scar

Krummel et al. hypothesized that TGF-beta, a profibrotic cytokine active during adult repair may similarly modulate the tissue response if added to fetal wounds [43]. Subcutaneous wound implants containing TGF-beta were placed in fetal rabbits at 24 days gestation, a time point when wounds heal scarlessly. When the implants were removed, histology showed marked fibroblast penetration and collagen deposition in contrast to the control fetal implants without TGF-beta. Likewise, in a human fetal skin repair model, addition of TGF-beta1 induced an adult-like inflammatory response in the fetal skin wound [63]. Human fetal skin was transplanted to a subcutaneous location on an adult athymic mouse and wounded. Wounded fetal skin in this environment heals scarlessly. However, when wounds were treated with a TGF-beta slow release disk, an adult-like inflammatory response was detected and the skin healed with scar formation. Thus, addition of cytokines with pro-fibrotic function to fetal wounds induces scar.

The lack of inflammation characteristic of fetal wounds, and the prominent inflammatory response in adult wounds, suggests inflammation itself may be a contributing factor to the scarring process. To study the capability of the fetus to respond to antigen and the changes in wound phenotype if inflammation was induced, sponge implants with lethally irradiated or live bacteria were placed subcutaneously in fetal rats. Upon harvesting the implants, histological sectioning revealed dose dependent acute inflammatory responses, confirming the ability of the fetus to generate an inflammatory response. Furthermore, there was increased collagen deposition in the bacteria-treated implants, which was most prominent in implants with live bacteria, implicating a role for inflammation in scarring [64]. To date, all experiments that increase inflammation during fetal repair have also induced scar formation.

Certain anti-inflammatory interleukins may play a role in promoting the scarless wound phenotype. IL-10 is a cytokine which decreases production of IL-6 and IL-8. Although it has not been added to adult wounds, the opposite experiment has been reported. Fetal skin grafted onto the back of adult IL-10 knockout mice heal incisional wounds with excess inflammation and scar [65].

## 6.2 Fibromodulin Promotes Scarless Fetal Repair

Fibromodulin is one of several TGF-beta modulators in the decorin proteoglycan family that can bind and inhibit TGF-beta activity. Based on previous experiments inversely correlating expression with scarring, Soo et al. used a fetal rat model to further explore the role of fibromodulin in fetal skin repair [66]. Exogenous fibromodulin was added to scarring fetal dermal wounds and resulted in hair follicle regeneration with decreased scar formation. Confocal microscopy revealed thin reticular collagen fibers similar to non-wounded age-matched skin. Furthermore, blocking of fibromodulin activity with anti-fibromodulin antibodies in fetal wounds that normally heal scarlessly induced a scarring pattern with thick disorganized collagen fibrils. This suggests fibromodulin may be involved in scarless dermal wound repair.

## 6.3 Smad3 or Hoxb13 Knockout Mice Exhibit Enhanced Wound Healing

Smad proteins are down-stream mediators of TGF-beta function, acting as nuclear transcription activators. TGF-beta signaling occurs through both Smad dependent and independent pathways. To deduce the targets of Smad signaling, animals lacking the Smad3 gene were generated [67]. Since TGF-beta is known to be intricately involved in adult wound healing, it was a surprise to find enhanced reepithelialization in Smad-null mice. This was associated with a reduced local inflammatory response. Therefore, disruption of the Smad3 pathway may be beneficial in wound healing.

Hoxb13 is a homeobox gene with decreased expression in scarless compared to scarring fetal wounds. Wounds in adult mice with null Hoxb13 expression heal with decreased scar and increased hyaluronic acid expression, giving a more "fetal-like" response [68].

## 6.4 Addition of TGF-Beta 3 or Blocking Antibody to TGF-Beta 1 and 2 Decreases Scar Formation

The TGF-beta isoforms have important roles in wound healing. TGF-beta 1 and 2 are profibrotic while TGF-beta 3 is thought to be anti-fibrotic. Shah et al. injected blocking antibody to TGF-beta 1 and 2 into open incisional wounds in adult rats [69]. Only addition of antibody to both isoforms decreased the inflammatory cells, neovascularization, and collagen deposition in the early stages of wound healing. In contrast, exogenous administration of TGF-beta 3 peptide produced similar decreases in inflammation and collagen deposition. Both treatments improved the architecture of the neodermis; however, no hair follicles formed in the wounds, thus true skin regeneration was not achieved. However, these and other studies have defined the pro-fibrotic function of TGF-beta 1 and TGF-beta 2, contrasting with the anti-fibrotic function of TGF-beta 3.

# 7 Perspective

Fetal wound repair is a complex sequence of events occurring superimposed on the tightly coordinated process of skin development. Disruption in the orderly sequence of regeneration with exogenous cytokines or inflammatory stimuli results in scar formation. Conversely, blockage of key profibrotic growth factors during adult repair produces a more fetal-like wound. To further dissect differences between scarless and scarring wounds, an advanced understanding of the effect of skin development on repair will be necessary. We are using microarray analysis to blueprint the scarless and scarring repair processes and provide insights for candidate genes that can be manipulated to enhance healing and regenerate skin. These genes may be manipulated in the adult wound to reduce scarring or in other organ systems to promote tissue regeneration.

## References


1. Byrne C, Hardman M, Nield K (2003) J Anat 202:113
2. Moore KL (1998) Before we are born: essentials of embryology and birth defects. Saunders, Philadelphia
3. Millar SE (2002) J Invest Dermatol 118:216
4. Longaker MT, Whitby DJ, Ferguson MW, Lorenz HP, Harrison MR, Adzick NS (1994) Ann Surg 219:65
5. Lorenz HP, Longaker MT (2003) J Craniofacial Surg 14:504
6. Longaker MT, Whitby DJ, Jennings RW, Duncan BW, Ferguson MW, Harrison MR, Adzick NS (1991) J Surg Res 50:375
7. Meuli M, Lorenz HP, Hedrick MH, Sullivan KM, Harrison MR, Adzick NS (1995) J Pediatr Surg 30:392
8. Mast BA, Albanese CT, Kapadia S (1998) Ann Plast Surg 41:140
9. Longaker MT, Burd DA, Gown AM, Yen TS, Jennings RW, Duncan BW, Harrison MR, Adzick NS (1991) J Pediatr Surg 26:942
10. Slate RK, Posnick JC, Wells MD, Goldstein JA, Keeley FW, Thorner PS (1993) Plast Reconstr Surg 92:874
11. Blewett CJ, Cilley RE, Ehrlich HP, Dillon PW, Blackburn JH II, Krummel TM (1995) J Pediatr Surg 30:945
12. Blewett CJ, Cilley RE, Ehrlich HP, Blackburn JH II, Dillon PW, Krummel TM (1997) J Thorac Cardiovasc Surg 113:880
13. Clark RA (1996) Molecular and cellular biology of wound repair. Plenum Press, New York
14. Merkel JR, DiPaolo BR, Hallock GG, Rice DC (1988) Proc Soc Exp Biol Med 187:493
15. Hallock GG, Merkel JR, Rice DC, DiPaolo BR (1993) Ann Plast Surg 30:239
16. Longaker MT, Whitby DJ, Adzick NS, Crombleholme TM, Langer JC, Duncan BW, Bradley SM, Stern R, Ferguson MW, Harrison MR (1990) J Pediatr Surg 25:63
17. Whitby DJ, Ferguson MW (1991) Development 112:651
18. Beanes SR, Hu FY, Soo C, Dang CM, Urata M, Ting K, Atkinson JB, Benhaim P, Hedrick MH, Lorenz HP (2002) Plast Reconstr Surg 109:160
19. Wang ZL, Inokuchi T, Ikeda H, Baba TT, Uehara M, Kamasaki N, Sano K, Nemoto TK, Taguchi T (2002) Int J Oral Maxillofac Surg 31:179
20. Lovvorn HN III, Cheung DT, Nimni ME, Perelman N, Estes JM, Adzick NS (1999) J Pediatr Surg 34:218
21. Mast BA, Flood LC, Haynes JH, DePalma RL, Cohen IK, Diegelmann RF, Krummel TM (1991) Matrix 11:63
22. Kennedy CI, Diegelmann RF, Haynes JH, Yager DR (2000) J Pediatr Surg 35:874
23. Clark RA (1996) Molecular and cellular biology of wound repair. Plenum Press, New York
24. Longaker MT, Whitby DJ, Ferguson MW, Harrison MR, Crombleholme TM, Langer JC, Cochrum KC, Verrier ED, Stern R (1989) J Pediatr Surg 24:799
25. Moulin V, Plamondon M (2002) Br J Dermatol 147:886
26. Beanes SR, Dang C, Soo C, Wang Y, Urata M, Ting K, Fonkalsrud EW, Benhaim P, Hedrick MH, Atkinson JB, Lorenz HP (2001) J Pediatr Surg 36:1666
27. Soo C, Hu FY, Zhang X, Wang Y, Beanes SR, Lorenz HP, Hedrick MH, Mackool RJ, Plaas A, Kim SJ, Longaker MT, Freymiller E, Ting K (2000) Am J Pathol 157:423

28. Dang CM, Beanes SR, Lee H, Zhang X, Soo C, Ting K (2003) Plast Reconstr Surg 111:2273
29. Huang EY, Wu H, Island ER, Chong SS, Warburton D, Anderson KD, Tuan TL (2002) Wound Repair Regen 10:387
30. Ihara S, Motobayashi Y, Nagao E, Kistler A (1990) Development 110:671
31. Lorenz HP, Lin RY, Longaker MT, Whitby DJ, Adzick NS (1995) Plast Reconstr Surg 96:1251
32. Cass DL, Bullard KM, Sylvester KG, Yang EY, Longaker MT, Adzick NS (1997) J Pediatr Surg 32:411
33. Lorenz HP, Whitby DJ, Longaker MT, Adzick NS (1993) Ann Surg 217:391
34. Olutoye OO, Yager DR, Cohen IK, Diegelmann RF (1996) J Pediatr Surg 31:91
35. Olutoye OO, Barone EJ, Yager DR, Cohen IK, Diegelmann RF (1997) J Pediatr Surg 32:827
36. Olutoye OO, Alaish SM, Carr ME Jr, Paik M, Yager DR, Cohen IK, Diegelmann RF (1995) J Pediatr Surg 30:1649
37. Olutoye OO, Barone EJ, Yager DR, Uchida T, Cohen IK, Diegelmann RF (1997) J Pediatr Surg 32:1037
38. Jennings RW, Adzick NS, Longaker MT, Duncan BW, Scheuenstuhl H, Hunt TK (1991) J Pediatr Surg 26:853
39. Clark RA (1996) Molecular and cellular biology of wound repair. Plenum Press, New York
40. Hildebrand A, Romaris M, Rasmussen LM, Heinegard D, Twardzik DR, Border WA, Ruoslahti E (1994) Biochem J 302(2):527
41. Scheid A, Wenger RH, Schaffer L, Camenisch I, Distler O, Ferenc A, Cristina H, Ryan HE, Johnson RS, Wagner KF, Stauffer UG, Bauer C, Gassmann M, Meuli M (2002) Faseb J 16:411
42. Nath RK, LaRegina M, Markham H, Ksander GA, Weeks PM (1994) J Pediatr Surg 29:416
43. Krummel TM, Michna BA, Thomas BL, Sporn MB, Nelson JM, Salzberg AM, Cohen IK, Diegelmann RF (1988) J Pediatr Surg 23:647
44. Dang C, Beanes SR, Soo C, Hedrick MH, Lorenz HP (2001) Wound Repair Regen 9:153
45. Soo C, Beanes SR, Hu FY, Zhang X, Dang C, Wang Y, Nishimura I, Freymiller E, Longaker M, Lorenz HP, Ting K (2004) Am J Pathol (in press)
46. Koo SH, Cunningham MC, Arabshahi B, Gruss JS, Grant JH III (2001) Plast Reconstr Surg 108:938
47. Whitby DJ, Ferguson MW (1991) Dev Biol 147:207
48. Dang CM, Beanes SR, Soo C, Ting K, Benhaim P, Hedrick MH, Lorenz HP (2003) Plast Reconstr Surg 111:1969
49. Beanes SR, Dang C, Soo C, Lorenz HP (2001) Wound Repair Regen 9:154
50. Liechty KW, Adzick NS, Crombleholme TM (2000) Cytokine 12:671
51. Liechty KW, Crombleholme TM, Cass DL, Martin B, Adzick NS (1998) J Surg Res 77:80
52. Lorenz HP, Adzick NS (1993) West J Med 159:350
53. Chen WY, Grant ME, Schor AM, Schor SL (1989) J Cell Sci 94(3):577
54. Schor SL, Grey AM, Ellis I, Schor AM, Howell A, Sloan P, Murphy R (1994) Cancer Treat Res 71:277
55. Ellis IR, Schor SL (1996) Exp Cell Res 228:326
56. Estes JM, Vande Berg JS, Adzick NS, MacGillivray TE, Desmouliere A, Gabbiani G (1994) Differentiation 56:173
57. Ravichandran KS (2001) Oncogene 20:6322
58. Ikeda K, Wang LH, Torres R, Zhao H, Olaso E, Eng FJ, Labrador P, Klein R, Lovett D, Yancopoulos GD, Friedman SL, Lin HC (2002) J Biol Chem 277:19206
59. Jung F, Haendeler J, Hoffmann J, Reissner A, Dernbach E, Zeiher AM, Dimmeler S (2002) Circ Res 91:38
60. Chin GS, Kim WJ, Lee TY, Liu W, Saadeh PB, Lee S, Levinson H, Gittes GK, Longaker MT (2000) Plast Reconstr Surg 105:972

61. Stelnicki EJ, Komuves LG, Holmes D, Clavin W, Harrison MR, Adzick NS, Largman C (1997) Differentiation 62:33
62. Stelnicki EJ, Arbeit J, Cass DL, Saner C, Harrison M, Largman C (1998) J Invest Dermatol 111:57
63. Lin RY, Sullivan KM, Argenta PA, Meuli M, Lorenz HP, Adzick NS (1995) Ann Surg 222:146
64. Frantz FW, Bettinger DA, Haynes JH, Johnson DE, Harvey KM, Dalton HP, Yager DR, Diegelmann RF, Cohen IK (1993) J Pediatr Surg 28:428
65. Liechty KW, Kim HB, Adzick NS, Crombleholme TM (2000) J Pediatr Surg 35:866
66. Soo C, Beanes SR, Dang C, Zhang X, Ting K (2001) Surg Forum 52:578
67. Liu X, Wen FQ, Kobayashi T, Abe S, Fang Q, Piek E, Bottinger EP, Roberts AB, Rennard SI (2003) Cell Motil Cytoskeleton 54:248
68. Mack JA, Abramson SR, Ben Y, Coffin JC, Rothrock JK, Maytin EV, Hascall VC, Largman C, Stelnicki EJ (2003) Faseb J 17:1352
69. Shah M, Foreman DM, Ferguson MW (1995) J Cell Sci 108(3):985

Received: December 2003

Adv Biochem Engin/Biotechnol (2005) 93: 101–134
DOI 10.1007/b99968


# Liver Regeneration

George K. Michalopoulos (✉) · Marie DeFrances

Department of Pathology, University of Pittsburgh School of Medicine, Pittsburgh, PA, 15241, USA
*michalopoulosgk@upmc.edu*
*defrancesmc@upmc.edu*



**Abstract** This review summarizes the functional aspects, cellular kinetics and molecular mechanisms related to liver regeneration. Liver regeneration is a model of regenerative growth of tissues in adult animals. Rapid biochemical and gene expression changes following initiation of regeneration are mediated by specific stimuli, including growth factors and cytokines. The whole process involves multiple feedback loops between same or different types of cells. The end result is restoration of hepatic mass and preservation of the normal histology of the liver. All of this complex phenomenology occurs while the liver continues to provide all the metabolic support required to sustain life of the organism.

**Keywords** Liver · Regeneration · Functional aspects · Cellular kinetics · Molecular mechanisms

**List of Abbreviations**

| | |
|---|---|
| A1R | Alpha-1 adrenergic receptor |
| AAF | *N*-2-Acetylaminofluorene |
| EGF | Epidermal Growth Factor |
| FAH | Fumaryloacetoacetate hydrolase |
| FGF | Fibroblast Growth Factor |
| HB-EGF | Heparin binding EGF |
| HGF | Hepatocyte Growth Factor |
| HGFL | HGF-like protein |
| IGF1 | Insulin-like Growth Factor 1 |
| IL6 | Interleukin 6 |
| MMP | Matrix Metalloproteinase |
| MT MPP | Membrane type MMP |
| PHx | 2/3 Partial hepatectomy |
| TACE | Tumor Necrosis alpha Converting Enzyme |
| TGF | Transforming Growth Factor |
| TNF | Tumor Necrosis Factor |
| uPA | Urokinase type plasminogen activator |
| VEGF | Vascular Endothelial Growth Factor |

# 1 Introduction

The liver performs many essential functions for the body, including regulation of plasma glucose levels (essential for bodily function); synthesis of many if not most of the proteins circulating in the plasma, from albumin to coagulation proteins; incorporation of toxic ammonia generated by transaminase reactions into the non-toxic urea; mobilization and export of lipids; synthesis and export to the intestine of bile, important for absorption of fat in the diet; storage of several vitamins; degradation of pigments, etc. Removal of the liver without providing metabolic support results in loss of life within a few hours. The liver is also the gateway of entry for all chemical substances ingested through the foods and absorbed through the gastrointestinal tract. The *portal vein*, formed by the tributary veins of the intestinal tract, the spleen and pancreas, bring to the liver all substances absorbed form the food as well as products (cytokines, hormones, etc.) from the spleen and pancreas. Most of the absorbed substances are compatible with cellular metabolic processing and can be utilized intact or can be further transformed into components of cells and tissues throughout the body. Many substances however cannot be further processed or incorporated into cellular structures. These substances, collectively called *xenobiotics*, enter the liver via the portal circulation and are further processed by *hepatocytes* (the main functional cells of the liver) so that they can be transported for excretion through the bile or the urine (after traveling through the circulation). Processing of xenobiotics entails interaction of the chemicals with mixed function oxygenases, such as members of the family of cytochrome P450 enzymes. The enzymatic reactions involved typically attach hydroxyl moieties to insoluble chemicals. The – OH groups become the sites for conjugation with other hydrophilic groups (sulfates, glucuronides, etc.) These reactions render the often insoluble and hydrophobic xenobiotics into soluble compounds amenable to transport and excretion. This overall detoxification process, however, often generates toxic free radicals of relatively long life, typically electrophilic, which can react with cellular nucleophiles such as DNA and proteins, causing hepatocyte damage and often cell death. Numerous studies in the decades of the 1970s and 1980s have defined the pathways involved in these processes by using specific model compounds such as acetaminophen, carbon tetrachloride, and a variety of chemical carcinogens [1, 2]. Sufficient quantities of such toxic xenobiotics can cause acute, severe, toxicity to the liver destroying most of the hepatocytes and causing death of the animal.

In addition to toxic xenobiotics, serious damage to the liver can be inflicted by several viruses. Both the viruses and the toxic xenobiotics can cause massive loss of hepatocytes (fulminant hepatitis), making liver functions unavailable and causing death of the organism. Long term toxic or inflammatory injury to the liver may also cause diffuse scarring of the liver. The scars run through the entire liver parenchyma and surround nodules of hepatic tissue. This condition is known as liver *cirrhosis*. The surviving hepatic tissue has abnormal vascular pathways resulting in hypoxia, which further confounds the effects of the toxic or inflammatory injury. Liver damage also occurs in most organisms living in their natural habitats. Ancient cultures performed animal sacrifices and used the scars on the surface of the liver (caused by such events and very frequently seen on the surface of animals living in the wild) as guides for the prediction of future events [3]. The wide abundance of such scars suggests that toxic injury to the liver of animals living in the wild is a relatively common event. The susceptibility of liver to massive histologic injury and the resultant danger to the survival of the organism, are perhaps the basis for the evolutionary forces that imparted liver with a large capacity to regenerate. Regeneration of liver was a key component of the Greek myth of Prometheus [3].

## 2 Histologic Aspects and Cellular Kinetics

Entering at the lower part of the liver, facing the abdominal cavity, are three major vessels. The *hepatic artery* (bringing oxygenated blood from the heart), the *portal vein* (carrying low oxygen but nutrient rich blood from the intestines, pancreas and spleen) and the *bile duct* (carrying bile out of the liver and into the intestine). These three vessels branch extensively throughout the liver but their small branches always maintain their proximity to each other, in the

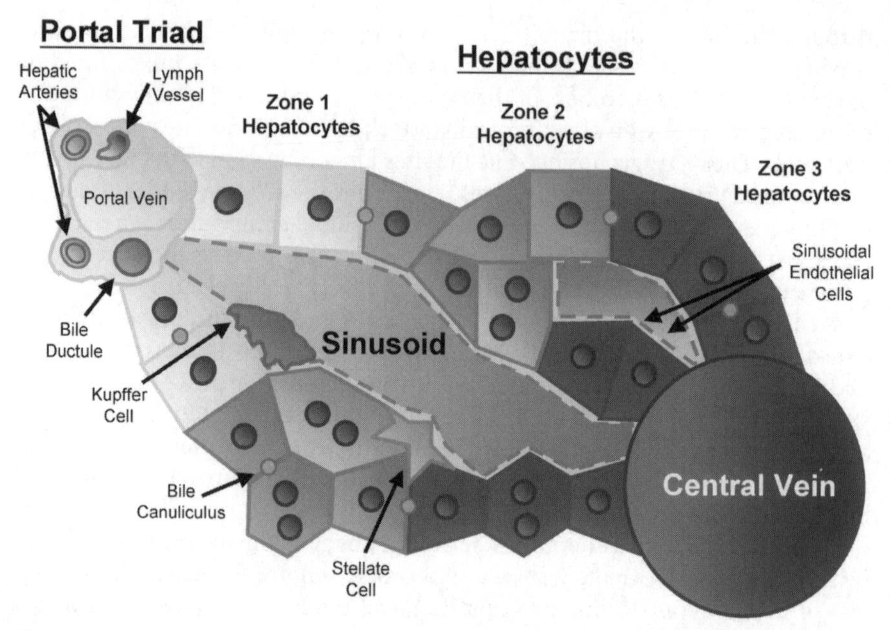


**Fig. 1** Diagrammatic outline of the key cellular elements and their location in the histologic architecture of the hepatic lobule

form of the *portal triads*. The triads are the hallmark of the histologic units of the liver called *lobules*. (A reverse approach centering around the central veins defines a histologic unit of acinus; also, please note that lobules are microscopic histologic units, in contrast to *lobes* which are large anatomical structures composed of thousands of lobules). Blood entering through the branches of the portal vein and the hepatic artery mixes and traverses through the hepatocytes. The latter are arranged into thin single or two cell plates, surrounded by endothelial cells which carry rather larger holes (fenestrae) in their cytoplasm and membranes. The *fenestrated endothelial cells* line the capillaries carrying blood on either side of the hepatocyte plates. These capillaries are called *sinusoids*. The sinusoids eventually merge into the *central veins* of the lobules, which merge into larger vessels and carry the blood back into the general circulation (Fig. 1).

Liver regeneration occurs every time there is major loss of hepatic tissue. This phenomenon occurs in all vertebrate organisms in which the liver exists as a distinct organ. Such loss of hepatic tissue may be easily induced by administration of hepatic toxins. CCl4 is a commonly used toxin for that purpose and it causes hepatocyte necrosis with a selective zonal pattern within the lobule, affecting primarily the pericentral areas. Most of the hepatic toxins cause necrosis with a similar zonal distribution, because hepatocytes in the pericentral areas have much higher expression levels of the enzymes involved in the biochemical transformation of the toxins to their reactive electrophiles [4].

The most commonly studied form of liver regeneration, however, is that of the restoration of hepatic mass following surgical removal of hepatic tissue. The model of this approach for the rat and the mouse was described by Higgins and Anderson [5]. Mouse and rat liver is composed of five lobes, each an intact histologic unit with portal vein, hepatic artery and bile duct branches. A simple surgical procedure (*partial hepatectomy* (*PHx*)) removes 70% of the liver mass (three of the five lobes). The residual 30% of the liver grows to restore the mass of the organ back to 100% of the original. It should be emphasized that the removal of the three lobes (from the total of five) does not incur any damage to the remaining two other lobes. Each lobe has its own vascular and bile duct supply, which is left intact. There is no wound to repair other than incision of the abdominal wall and the ligated vascular stalks of the three removed lobes. It should be emphasized that this growth does not restore the shape of the liver. The restored liver is now composed of two lobes, instead of five. In addition, this growth does not create new lobules. Rather the size of the individual lobules increases and the hepatocyte plates of the regenerated liver have a thickness of two cells on the average, compared to the single cell thickness prior to the operation. Similar techniques are used in human liver surgery, though human liver is a single piece of tissue (functionally subdivided into two lobes). Hepatectomies in humans leave an open wound through the hepatic parenchyma that needs to heal in a separate fashion. Though the growth is restores the total liver mass, the term "regeneration" (typically implying restoration of both mass and shape) is used to describe it. This process of liver regeneration is rapid, well coordinated, and in many ways spectacular. Liver function is maintained with very few changes during the entire regenerative process, including all biosynthetic functions, detoxification pathways, etc.

PHx is associated with many rapid changes in all cellular populations of the liver. In addition to *hepatocytes*, other cell populations of importance are sinusoidal *endothelial* cells, *biliary epithelial* cells, and *stellate* cells. The latter are highly specialized cell types located between the endothelial cells and the hepatocytes [6]. They have long processes engulfing hepatocytes. They synthesize extracellular matrix proteins, store vitamin A, produce growth factors and have a gene expression pattern very similar to the astrocytes of the brain. Hepatocytes first enter into DNA synthesis. A measurable peak of DNA synthesis is seen in hepatocytes at 24 h in the rat and at 36 h in the mouse, following PHx. The course in humans is more variable, depending on nutritional status, other underlying diseases, etc. [3].

During the first wave of proliferation, most of the hepatocytes enter into DNA synthesis, though not at the same time. It should be emphasized at this point that liver regeneration is mediated by participation of most of the hepatocytes, and that it is not dependent on proliferation of a select sub-population with stem cell characteristics. Detail studies involving thymidine incorporation have demonstrated that more than 90% of the hepatocytes participate in the regenerative process [3]. Hepatocyte proliferation starts at the sites around the portal triads (*periportal* hepatocytes), and proliferation progresses and reaches the hepatocytes located around the central veins (*pericentral* hepatocytes) by about 48 h in the rat [7]. Proliferation of the other cells of the liver lobule proceeds at different paces. Stellate cells and biliary epithelium reach a first proliferative peak at 48 h and proliferate slowly thereafter. Endothelial cells start proliferating at three days with a peak at five days after PHx [8].

## 3 Growth Regulation Networks Between Different Hepatic Cell Populations

The process of regeneration aims to replace lost tissue. Thus, proliferation of hepatocytes alone is not sufficient. All other cells participate. Hepatocytes produce multiple growth regulatory cytokines (Fig. 2) which have known stimulatory effects on growth of the adjacent cells. Growth factors produced by the hepatocytes include TGFα, VEGF, FGF1, FGF2, Angiopoietins 1 and 2, PDGF, HB-EGF, etc (see below). All of these factors can stimulate growth of endothelial cells, stellate cells and biliary epithelium. Some of them (e.g., TGFα, FGF1, FGF2, HB-EGF) may also stimulate growth of hepatocytes. The possibility of autocrine stimulatory effects of the factors produced by hepatocytes on hepatocytes themselves has not been firmly established but cannot be ruled out. Growth factors affecting hepatocytes are also produced by the adjacent stellate and endothelial cells. HGF is produced by the adjacent stellate cells [9] and also by endothelial cells [10]. Recent evidence suggests that VEGF produced by the replicating hepatocytes stimulates production of the hepatocyte mitogen HGF by the endothelial cells via the VEGF receptor 1 [10]. Same interacting networks also operate between HGF and TGFα. Administration of HGF to hepatocytes in culture leads to production of TGFα, a mitogen for hepatocytes, endothelial cells and stellate cells [11]. IL6, produced by Kupffer cells and hepatocytes, stimulates acute phase synthesis and accelerates activation of the transcription factor Stat-3 [12]. IL6 is involved in "priming" of hepatocytes in order to render them sensitive to mitogenic of growth factors (see below). IL6 is also a direct mitogen for biliary epithelial cells [13]. Stellate cells also synthesize TGFβ1. This cytokine, although a mitoinhibitor for hepatocytes in culture, does not affect proliferation of hepatocytes during regeneration. It does stimulate, however, hepatocyte motogenesis (as does HGF) and, more importantly, it is a regulator for production of extracellular matrix by responding cells (see below). Recent evidence from organoid cultures shows that HGF and EGF (the latter arriving to the liver from glands of the gastrointestinal tract) stimulate production of TGFβ1 in stellate cells [14, 15]. Norepinephrine, another substance involved in enhancing effects of mitogens, also stimulates production of new HGF and EGF from source tissues (see 5G). Notably, similar pathways have been described in most malignant neoplastic cells. The evidence from growth regulation of normal tissues, such as liver, suggests that the paracrine growth regulatory effects of the neoplastic cells mimic the pathways operating in the normal tissues from which the neoplasms arise.

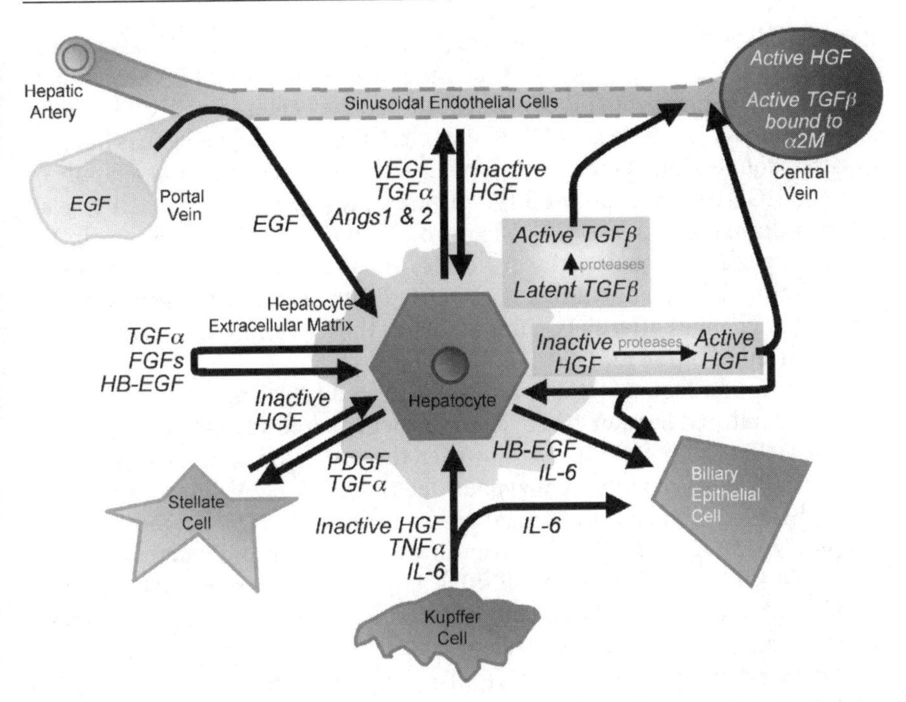


**Fig. 2** Interacting networks of signaling growth factors and cytokines produced by hepatocytes and the other cellular elements of the liver during liver regeneration

## 4
## Changes in Biochemical Properties and Gene Expression Patterns Associated with Liver Regeneration

PHx triggers a series of changes that occur very soon after hepatic tissue is removed. Urokinase (uPA) activity is upregulated within minutes, apparently affecting the entire hepatic parenchyma [16]. The reasons are not entirely clear, but they may be related to transport of the uPA receptor (uPAR) to the plasma membrane, thus making it available to bind and upregulate uPA activity. Hepatocyte membrane is rapidly depolarized [17]. This persists for more than 3 h after PHx and it may correlate with the increased levels of GABA (gamma aminobutyric acid) in the plasma. GABA is present in the liver at high concentrations. Interference with the hepatocyte membrane depolarization by GABA antagonists, however, has no apparent effect on the early changes in gene expression seen after PHx [18].

The receptors for epidermal growth factor (EGF) and hepatocyte growth factor (HGF) are activated by tyrosine phosphorylation within 30–60 min, presumably through increased binding to their ligands [19]. Alternatively, one of the receptors may be activated and then cross-activate the other by receptor cross-talk, a phenomenon that has been described in hepatocyte-derived neoplastic cell lines [20]. A cluster of genes, ("immediate early genes") is also rapidly upregulated [21–23]. Many of these genes are related to cell cycle and have been shown to increase in other cell types when they enter the cell cycle. These genes include c-Jun, c-Fos, c-Myc, K-Ras, etc. Other upregulated genes have no apparent connection to the cell cycle events. IGF-BP1, a member of a family of proteins (IGF-BP) that can bind the insulin-like growth factor 1 (IGF1), also shows a dramatic increase [24]. This protein may also play a role for intracellular signaling related to the cell cycle, independent of the IGF1 binding. Homozygous deletion of IGFBP1 is associated with delayed DNA synthesis and pathologic changes after PHx comparable to those seen in mice with homozygous deletion of IL6 [25].

Transcription factor Stat-3 is rapidly activated [26]. This response can be elicited in cultured hepatocytes by many growth factors and cytokines, including IL6, HGF, etc. [27, 28]. In vivo, however, the activation of stat3 is decreased dramatically in mice with homozygous deletion of IL6 [29]. Surprisingly, this is not observed in mice with homozygous deletion of the IL6-shared common receptor gp130 [30, 31]. Another transcription factor rapidly activated after PHx is NF-kB. The activation of NF-kB is dependent upon removal of an associated inhibitory protein (IkB). The pathways controlling this process in many cell types have been extensively reviewed [32]. Relevant to liver regeneration is the fact that deficient activation of NF-kB is observed in mice with homozygous deletion of the tumor necrosis factor receptor 1 (TNFR1), suggesting that this cytokine (TNF) is responsible for the proper timing of NF-kB activation during the regenerative process. Mice with deficient expression in TNFR1 also have deficient liver regeneration [33, 34]. Given the fact that hepatic regeneration eventually proceeds in both IL6 -/- and TNFR1 -/- mice, activation of Stat-3 and NF-kB must be compensated by other growth factors and cytokines in the absence of the IL6 and TNF triggered signaling cascades. (For more discussion on NF-kB see below).

Many intracellular signaling cascades involve pathways described in most cell types during proliferation, involving activation of mitogen activated protein kinase (MAPK), signaling dependent on IP3 kinase, etc. Beta catenin, a protein involved in regulation of expression of many cell cycle proteins, increases transiently within the first 30 min after PHx [35]. This transient increase in beta catenin is followed by rapid increase of APC, a protein forming part of a complex that can phosphorylate beta catenin in serine and threonine sites and thus tag it for subsequent ubiquitination and degradation by proteasomes. The increase in APC reflects efforts by cells to limit increase of beta catenin to a manageable short-term pulse. Increase in free cytoplasmic beta catenin is followed by rapid translocation to the nucleus of hepatocytes. Increased levels of beta catenin can be demonstrated in the nuclei for the first 2–24 h after PHx [35]. Recent studies have shown that the increase in beta catenin, in addition to events mediated by the Wnt/frizzled system, may also be due to tyrosine phosphorylation of beta catenin by receptor tyrosine kinases such as the HGF and EGF receptor [36]. Point mutations in beta catenin have been described in hepatocellular carcinomas [37–39]. The involvement of beta catenin in the early events after PHx provides a framework of understanding to explain the role of beta catenin in liver carcinogenesis. Many of the point mutations described eliminate serine- and threonine-phosphorylation of beta catenin, thus preventing its degradation. This allows beta catenin to increase in the cytoplasm and move to the nucleus, where it dimerizes with members of the Tcf family of transcription factors and induces gene expression of many cell cycle proteins including c-Myc, a protein upregulated in many hepatocellular carcinomas [40].

Despite the complexity of events occurring in hepatocytes during regeneration, there are very few noticeable decreases in indices of hepatic function. Levels of key plasma proteins produced by hepatocytes (albumin, coagulation factors, anti-proteases, etc.) are minimally changed, if at all. Accumulation of lipid droplets in hepatocytes cytoplasm is seen in the first 48 h after PHx. This is associated with increase in fatty acid and lipid synthesizing enzymes [41]. Marked changes occur in intercellular junctions associated with bile canaliculi [42]. These subcellular organelles are changing in structure and assume morphologic features seen in embryonic liver. All of the above changes are transient. Intercellular junctions and canalicular structures are completely restored by 72–96 h after PHx.

As with other tissues, liver regeneration is associated with remodeling of extracellular matrix [43]. This proceeds through two major pathways. The first pathway is associated with activation of urokinase. The latter causes enhanced activation of plasminogen to plasmin. Plasmin, in addition to its fibrinolytic activities, activates (via proteolytic cleavage) a variety of matrix related metalloproteinases. MMP9 (matrix metalloproteinase 9) is the one mostly activated during liver regeneration [44, 45]. Demonstrable activation of MMP9 is seen at 2–4 h after PHx. MMP9 protein can be demonstrated by immunohistochemistry in periportal hepatocytes at 3 h after PHx. The immunoreactivity gradually expands into the periportal and midzonal areas of the lobule. By 48 h it is limited to hepatocytes of the pericentral areas. This pattern follows that seen with hepatocyte proliferation and mitosis. The second pathway is mediated through a family of membrane bound metalloproteinases, MT MMP (membrane type metalloproteinase). Of the several members of this family, MT1 MMP exists in the liver [46]. MT1 MMP, acting together with TIMP2 (tissue inhibitor of metalloproteinase 2) activates MMP2. This pathway proceeds from 12 to 48 h after PHx. Changes in several extracellular matrix proteins are seen soon after PHx, including fibronectin, laminin, entactin [44]. Several groups have described deficient liver regeneration in mice with homozygous deletions of enzymes involved in matrix remodeling, including urokinase [47] and plasminogen [48]. In addition to problems with matrix remodeling, urokinase-deficient mice may have impaired regeneration due to deficient activation of HGF by urokinase (see below). Matrix remodeling is involved with tissue regeneration and cell proliferation in many systems and tissues, normal and neoplastic [49]. Pathways affected include signaling via integrins [50]; generation of proteolytic fragments with direct effects on receptor tyrosine kinases (most notable the EGF receptor [51]); and changes in levels of extracellular matrix components necessary for activation of receptors of the FGF family of growth factors [52]. Both HGF and EGF bind to the extracellular matrix of the periportal areas. Matrix remodeling mediated by urokinase causes release (and, in the case of HGF, also activation: see below) of important mitogenic growth factors for hepatocytes, thus potentially contributing to generation of mitogenic signaling cascades leading to hepatocyte proliferation. TGFβ1, a mitoinhibitory cytokine, is bound to decorin, an extracellular matrix protein attached to the plasma membrane [53]. Remodeling of the extracellular matrix is probably the cause for the rapid increase in plasma levels of TGFβ1 after PHx (see below). The potential signaling pathways resulting from extracellular matrix remodeling are quite complex. Overall, addition of complex matrix to hepatocyte cultures (such as type 1 collagen gels or Matrigel, a matrix product extracted from a mouse sarcoma [54]) inhibits hepatocyte proliferation in monolayer cultures [55]. Disruption of the hepatic extracellular matrix by perfusion of the liver with collagenase causes entry of hepatocytes into the cell cycle with enhanced expression of k-Ras, c-Myc and other cell cycle related proteins [56]. It also enhances effects of HGF administered in vivo, following collagenase infusion into the liver of live rats through the portal circulation [57].

## 5 Role and Effects of Individual Growth Factors and Cytokines in Triggering Liver Regeneration

### 5.1 Epidermal Growth Factor (EGF), Its Receptor (EGFR) and Associated EGFR Ligands

EGF is produced by many cells and tissues [58]. Its receptor (EGFR), a transmembrane receptor tyrosine kinase is present on most cells, including epithelial, mesenchymal and neuronal cells [59]. The final form of EGF is a 6-kD polypeptide derived from cleavage of a larger precursor. Though EGF precursor is found in many tissues (including brain and kidney), the mature form of the 6-kD polypeptide is secreted primarily by exocrine glands, including salivary glands (the primary site from which it was isolated) and exocrine glands of the gastrointestinal tract (including Brunner glands of the duodenum [60]). The salivary glands also secrete EGF into the plasma, in measurable concentrations. EGF is a direct mitogen for hepatocytes in primary monolayer cultures [61]. Infusion of EGF into liver through the portal vein results in deposition of EGF to the periportal extracellular matrix [62]. EGF infused into the portal vein is cleared by liver in one pass [62]. In addition, infusion of EGF into the liver causes DNA synthesis in periportal hepatocytes [63, 64]. In male mice, removal of the salivary glands (a major site of production of EGF) causes deficient liver regeneration [65, 66]. There is no increase in plasma EGF levels following PHx. There is, however, enhanced tyrosine phosphorylation of he EGF receptor within 30 min after PHx with a peak seen at 60 min [19]. It is possible that this activation may be due to crosstalk with the HGF receptor, since the timing of the activation of the two receptors is identical [20]. The strong mitogenic effects of EGF on hepatocytes in monolayer cultures and in whole animals and the rapid activation of EGFR prior to the emergence of other EGFR ligands (such as TGFα) strongly suggests that EGF reaching the liver through the portal circulation contributes to the mitogenic signaling events during the early stages of liver regeneration.

Another ligand of EGFR associated with liver regeneration is TGFα (transforming growth factor alpha). TGFα is derived via proteolytic cleavage from a larger precursor form which has a transmembrane domain [67], by proteolytic cleavage mediated by TACE [68]. The portion of the precursor molecule containing the mature form is located in the extracellular domain. The precursor form of TGFα has effects on adjacent cells expressing EGFR [69]. TGFα acts through binding on the EGFR. Though they share receptors, the effects of EGF and TGFα are not identical on most cellular targets [70], with the effects of TGFα tending to be more pronounced. The reasons for this are not clear and they appear to relate to differential processing of EGFR after binding to its ligands [71, 72]. TGFα is produced by most replicating epithelial cells, including hepatocytes in culture and during liver regeneration. TGFα mRNA increases in hepatocytes starting at 3–5 h after PHx and peaking at 24 h. Since hepatocytes express TGFα and its receptor EGFR, it is possible that the joint expression of both the ligand and the receptor may result in an autocrine mitogenic loop [73]. Since most hepatic cell types adjacent to hepatocytes also express EGFR, it is also likely that production of TGFα by hepatocytes may be involved in generation of mitogenic signals for the adjacent cells (stellate and endothelial cells). Direct infusion of TGFα to the liver enhances DNA synthesis in hepatocytes of normal rats [74]. Overexpression of TGFα under the albumin promoter induces a hyperproliferative phenotype in hepatocytes and the mice develop hepatocellular carcinomas [75, 76]. These studies suggest a key role or TGFα in liver regeneration. Other studies suggest however that this role, though of a key nature, may not be indispensable. Despite the dramatic increase in TGFα mRNA, the overall increase in the protein is rather small [77]. Mice with homozygous deletion of the TGFα gene have few phenotypic effects overall, suggesting that other ligands of EGFR (which include EGF, TGFα, Cripto, Amphiregulin, Epiregulin, etc.) may compensate for the loss of TGFα [78]. The TGFα deficient mice have normal liver regeneration, suggesting that the role of TGFα in liver regeneration may not be crucial or may be compensated by the other EGFR ligands [79].

Though the role of TGFα appears dispensable, the dramatic rise in its mRNA and the strong mitogenic effects on all hepatic cells types suggest that under normal circumstances following PHx, TGFα is a key mitogenic ligand for EGFR on all cell types expressing EGFR. Other EGFR ligands are also mitogenic in hepatocyte cultures. Amphiregulin is expressed in organoid liver cultures in which hepatocytes exhibit spontaneous growth. Its expression is suppressed by dexamethasone [15]. Heparin binding EGF (another ligand of EGFR) expression increases after PHx and transgenic mice with expression of HB EGF in hepatocytes have enhanced regenerative response [80–83]. Epiregulin is a strong mitogen for hepatocytes in culture, surpassing the effects of EGF and TGFα [84]. The above studies document that, though most of the studies have concentrated on EGF and TGFα, all ligands of EGFR may be playing distinct roles either in liver regeneration or in embryogenesis. Despite the above findings, however, mice with homozygous deletion of EGFR have a range of abnormalities in many tissues but not in liver [85]. Regeneration of the liver in EGFR deficient mice has not been studied. EGFR is a member of the Erb family of receptor tyrosine kinases. Erb2 is only seen in fetal liver and hepatocellular carcinomas. EGFR and Erb3 are the only members of this family of receptors seen in adult liver. The role of Erb3 in regeneration has not been fully characterized [86–88].

## 5.2 Hepatocyte Growth Factor (HGF) and Its Receptor (cMet)

This growth factor was isolated from plasma of rats subjected to partial hepatectomy [89, 90] as well as from human placenta [91]. Previous studies had shown that, following partial hepatectomy in the rat, transplanted hepatic fragments and transplanted isolated hepatocytes responded to PHx along with the liver of the host [92, 93]. Other studies had shown that PHx in one member of a pair of rats with parabiotic (joint) circulation resulted in a regenerative response to both members of the parabiotic pair [94]. The mitogenic principle was isolated based on its capacity to induce DNA synthesis in hepatocyte cultures kept in chemically defined media and was soon named hepatocyte growth factor (HGF). Notably, numerous other cytokines that appear to influence liver regeneration also rise in the plasma after PHx, including IL6, TNF, norepinephrine, TGFβ1, etc. HGF, however, is the only one circulating in the plasma that has direct mitogenic effects on hepatocytes in primary culture. Other cytokines increasing after PHx and relevant to liver regeneration do not have direct *mitogenic* effects on hepatocytes (see below). The HGF receptor was also identified soon after cloning of HGF as the product of the proto-oncogene c-met [95, 96]. Scatter Factor, a *motogenic* cytokine (induces cell motility), was also found to be identical to HGF [97]. HGF is also often referred as HGF/SF based on that discovery.

There are multiple reviews on the properties and functions of HGF and its receptor [98–100]. Following is only a summary of the general properties of HGF and its receptor that are relevant to this review:

1. HGF is a multifunctional cytokine produced primarily by cells of mesenchymal derivation, primarily fibroblasts [101], stellate cells of the liver [9],

endothelial cells [10] and some groups of neurons [102]. cMet, a transmembrane receptor tyrosine kinase, is expressed mostly in epithelial cells of most viscera and skin, in some groups of neurons and also in mesenchymal cells [99, 103].
2. Binding of HGF to its receptor induces mitogenic and motogenic effects, resulting in morphogenesis [104, 105].
3. The HGF molecule is composed of a heavy chain (containing an amino terminal hairpin loop and four kringle domains) and a light chain with a structure of pseudo protease. There is extensive amino acid sequence homology with plasminogen [91]. Another protein (known as HGF like protein or Macrophage Stimulating Protein also has a similar structure but different effects [106]. HGFL is produced by hepatocytes [107]. The receptor of HGFL (cRon) has extensive amino acid sequence homology and structure with cMet [108].
4. HGF is synthesized as a single chain inactive protein which binds to extracellular matrix sites. High concentrations of single chain HGF are present in the extracellular matrix of liver and placenta [109, 110]. The active heterodimeric form is produced via endoproteolytic cleavage at an Arg-Val-Val site. This site is identical to that used for proteolytic activation of plasminogen. Urokinase plasminogen activator (uPA) can activate HGF in test tube solutions and in vivo [16, 111]. Deficient activation of HGF in seen in uPA deficient mice, both in liver [112] and in the central nervous system [113]. Another protease, known as HGFA (HGF activator) can also activate HGF [114]. Other proteases have also been described to activate HGF, including plasmin, cathepsins, etc., albeit with lesser affinity and effectiveness.
5. Smaller variants of HGF containing only the amino terminal loop plus one or two of the first kringles have been described [115] and are thought to be derived from splicing variation. They are less active and require heparin for enhanced activity [116].
6. HGF upregulates expression of several anti-apoptotic proteins associated with signal transduction, such as Bcl Xl [117] in many cell types including hepatocytes. cMet, when not bound by HGF, dimerizes with the apoptotic receptor Fas and prevents it from forming trimers required for Fas activation and induction of apoptosis [118]. High levels of HGF may massively displace Fas from cMet and thus facilitate hepatocyte apoptosis. This may be of relevance in fulminant hepatic failure, in which there is massive hepatocyte apoptosis, high rates of hepatocyte DNA synthesis, very high levels of HGF and failure of the liver to regenerate [119–123].

HGF appears to be of key importance in relation to liver regeneration. The main lines of evidence linking HGF to liver regeneration are as follows:

1. HGF concentration rises 20-fold in the plasma shortly after PHx in rats [124] and humans [125]. Elevated activity persists for more than 48 h. It should be noted, however, that HGF also rises in plasma, albeit at lower levels, following injury to kidney, intestine, lung and other tissues [126–128].

2. uPA activity, capable of activating HGF, rises rapidly after PHx (see above). Addition of anti-uPA antibodies to whole tissue homogenates from regenerating liver blocks HGF activation [16]. Mice deficient in uPA also have deficient liver regeneration [47, 112].
3. The HGF receptor is rapidly activated by tyrosine phosphorylation within 30 min after PHx [19].
4. HGF mRNA increases in regenerating liver, starting at 3 h after PHx with peak levels reached by 24 h and remains elevated for the first 72 h [129]. Similar changes are seen in other HGF producing tissues, such as kidney and lung [130]. The message(s) responsible for the systemic elevation in HGF production are not clear. Substances known to stimulate HGF production in mesenchymal cells and also rising early after PHx include norepinephrine [131] and IL1a [132] and IL6 [133] and they may be involved in mediating this response. HGF production is also stimulated in endothelial cells by members of the vascular endothelial growth factor (VEGF) via the VEGF Receptor 1 [10].
5. Both single chain and two chain (active) HGF proteins decrease in liver for the first 3 h after PHx, suggesting consumption from the stores in the hepatic extracellular matrix. At 3 h, and coincidental with increase in HGF mRNA, there is rise of both single chain and active HGF, with a peak at 24–48 h [134].
6. Infusion of HGF through the portal or the systemic circulation in normal mice and rats causes massive hepatic enlargement [135]. Increase in liver weight and hepatocyte proliferation is also seen after infusion of plasmids which result in HGF expression [136].
7. Antibodies against HGF delay or suppress hepatocyte proliferation after PHx [137].
8. Mice with homozygous deletions of HGF and its receptor cMet die in utero by day 15 of embryonic development. Associated abnormalities include placental malformations, smaller liver size and decreased migration of myoblasts to peripheral muscle sites [138–140].
9. Active (two chain) HGF is one of the two strong mitogens for hepatocytes in culture (the other being ligands of EGFR, such as TGF$\alpha$, EGF and Amphiregulin). Both HGF and EGFR ligands provide complete mitogenic signals leading to hepatocyte proliferation in chemically defined (serum free) media. The effects of HF and EGF are synergistic. Together they provide the strongest mitogenic stimulus known for hepatocytes [141].

The combined evidence suggests that the rapid extracellular matrix remodeling following PHx causes local release and activation of HGF bound to extracellular matrix. This contributes to the signaling cascades leading hepatocytes to proliferation during the first 3 h after PHx. The HGF rise in the peripheral blood is probably related to the release and activation of HGF from hepatic extracellular matrix stores. It should be noted that the form of HGF rising in the plasma is that of the active (two-chain) HGF, further documenting that HGF is not only released abut also activated [134]. The role of HGF produced by de novo synthesis starting at 3 h after PHx is not clear. It is also not clear whether HGF synthesized in lung and kidney contributes to the observed increase in HGF protein in liver tissue during regeneration. The newly synthesized HGF is produced by stellate cells [9, 142]. Recent studies have also shown that VEGF stimulates production of HGF by endothelial cells through VEGFR 1. Hepatocytes do produce VEGF during liver regeneration [10]. This may be a stimulus responsible for the observed increase in HGF mRNA in liver during regeneration.

Several recent studies with hepatocytes in culture have shown that some of the effects of HGF may be mediated by induction of TGFα synthesis in hepatocytes. Blocking TGFα synthesis by anti-sense mRNA decreased the mitogenic effects of HGF [11]. There was similar decrease in HGF mitogenic effects by specific inhibitors of EGFR, the receptor for TGFα [143]. It is not clear whether these findings are specific to HGF or also to EGF itself. Most epithelial cell mitogens, including HGF, EGF, PDGF, FGF1, FGF2, etc. induce synthesis of new TGFα (in breast, intestine, kidney, and other tissues). TGFα precursor is a transmembrane protein and as such it may have many effects in cell structure and function which may not necessarily relate to its mitogenic effects. This finding needs to be further pursued.

In addition to its effects and changes during liver regeneration, HGF also has ameliorating effects in many types of chronic liver injury including liver cirrhosis [144, 145]. Analogous effects of HGF on chronic and acute tissue injury have been shown for other tissues such as lung and kidney [130].

## 5.3 Fibroblast Growth Factors (FGF)

This is a family of growth factors which encompasses 23 family members. Most of the members of this group have high affinity for glycosaminoglycans and often require presence of heparin or similar molecules in culture in order to express activity [146, 147]. Their effects are mitogenic for several cell types, including hepatocytes [148]. Both FGF1 and FGF2 (also known as acidic and basic FGF) are mitogenic in hepatocyte cultures, although less so than either HGF or EGF and TGFα. Their effects (as with most of the 22 members of the FGF family) are mediated through four fibroblast growth factor receptors (FGFR1–4) [149]. These receptors constitute a family of transmembrane tyrosine kinases. Increase in gene expression of FGF1 and FGF2 have been reported during liver regeneration [150, 151]. Mice deficient in FRGF1 and FGF2 or the FGFR4, however, have normal liver regeneration following PHx or administration of hepatotoxins such as CCl4 [151, 152]. It is highly likely that the absence of specific ligands or receptors in this growth factor family is compensated by the other members of the family. While the role of FGF family members in regeneration is not clear, studies by Zaret et al. have demonstrated that FGF1 is released by cardiac mesoderm during embryogenesis. In combination with members of the bone morphogenetic protein family (BMP), it leads ventral endoderm to form the hepatic primordium during embryogenesis [153, 154]. KGF (FGF7) is also mitogenic for hepatocytes in cell culture and in vivo [155, 156]. Overexpression of KGF in liver during development result in enlarged liver and multiple abnormalities in kidneys.

## 5.4 Stem Cell Factor and c-Kit

Recent studies have implicated stem cell factor (a ligand of the receptor c-Kit [157]) as playing a role in liver regeneration. Hepatocyte proliferation after PHx is decreased in SCF deficient mice and administration of SCF to IL6 deficient mice enhances hepatocyte proliferation (see below) [158]. SCF and its receptor c-Kit is also expressed in populations of oval cells growing in the AAF-suppressed liver regeneration after PHx [159] (see below).

## 5.5 Growth Factors Associated with Angiogenesis

Restoration of normal histology requires formation of the vascular network proving nourishment to hepatocytes and stellate cells. Angiogenesis is very precisely regulated during liver regeneration. Proliferation of the sinusoidal endothelial cells starts at day 2–3 after PHx. Following hepatocyte proliferation in the first 48 h, the newly formed hepatocytes are not as intensely vascularized as normal liver. Poorly vascularized hepatocyte clumps have been described [160] and can be seen by scanning electron microscopy [161]. Replicating endothelial cells invade the poorly vascularized hepatocyte clusters, permeate them and form new sinusoids, completing the restoration of the hepatic parenchyma. Proliferation of the endothelial cells is probably stimulated by a wide array of angiogenic factors produced by hepatocytes during regeneration. These include vascular endothelial growth factor (VEGF; it has multiple splicing variants) [162, 163], angiopoietins 1 and 2 [164, 165] as well as other growth factors which have angiogenic effects such TGFα, FGF1, FGF2. HGF, produced by the stellate cells (and by the endothelial cells themselves) also has angiogenic properties [166]. Angiopoietin receptors Tie-1 and Tie-2 are expressed in sinusoidal endothelial cells during regeneration [161, 167]. Of the VEGF receptors, Flk1 is expressed primarily in the endothelium of large vessels, whereas Flt1 is expressed in sinusoidal endothelium but also on hepatocytes [161]. Angiostatin, an inhibitor of angiogenesis, delayed and suppressed liver regeneration in mice [168]. Recent studies have also shown (see above) that VEGF can stimulate production of HGF in endothelial cells via the VEGF R1 (Flk1). This establishes a positive feedback proliferative loop between hepatocytes and endothelial cells, as proliferating hepatocytes produce VEGF which in turn stimulates production of HGF. On the other hand, proliferation of hepatocytes after PHx in the rat is mostly concluded by 48 h, whereas proliferation of endothelial cells starts at about the same point in time [8].

## 5.6 Tumor Necrosis Factor (TNF) and Interleukin 6 (IL6)

Several studies have implicated these two cytokines as playing a role in liver regeneration. In normal liver, TNF is produced mainly by the hepatic macrophages (Kupffer cells) [169]. TNF can stimulate production of IL6 in hepatocytes as well as in Kupffer cells [34, 170]. Antibodies against TNF decrease the regenerative response after PHx and inhibit activation of Jun Kinase [171]. TNF levels in plasma rise after PHx [172]. TNF itself is not mitogenic to hepatocytes in culture but augments the mitogenic effects of HGF and TGFα both in hepatocyte cultures and in vivo, following HGF or TGFα injections [74]. Mice deficient in TNF receptor 1 (TNFR1) have deficient liver regeneration [33, 34, 133]. TNF, as the name implies, most often induces apoptosis on many cell types, especially tumor cells. It may however induce pro-mitogenic effects, and that depends on whether TNF induces activation of NF-kB [173]. The transcription factor NF-kB is normally bound to the protein IkB, which prevents NF-kB activation. Phosphorylation of IkB causes disassociation from NF-kB and allows NF-kB to travel to the nucleus and affect transcription of several genes associated with cell replication. Mice with homozygous deletion of one of the NF-kB subunits (p65) die in utero with embryonic liver failure and massive hepatocyte apoptosis [174]. NF-kB is rapidly activated following PHx [175]. This event is dramatically decreased in mice deficient in TNFR1, which also exhibit deficient liver regeneration [34]. This led to the hypothesis that TNF mediated activation of NF-kB is a key event in liver regeneration and that TNF acts as a mitogenic factor in vivo, causing "priming" of hepatocytes and enhancing their sensitivity to direct mitogens, such as HGF, EGF and TGFα. Several findings have reinforced this hypothesis. A deletion mutant form of IkB lacks the phosphorylation site mediating its disassociation from NF-kB and it is a constitutive NF-kB inhibitor, known as "super-IkB". When adenoviral constructs forcing expression of super-IkB were injected to the liver, PHx did not result in liver regeneration and was followed with massive hepatocyte apoptosis [176]. The analysis of the results, however, was complicated by the fact that the adenovirus vector itself also caused hepatocyte apoptosis, induction of interferon and release of TNF and IL6. In more recent studies, controlled expression of super-IkB in hepatocytes was achieved using a controlled expression double transgenic mouse strain, in which super-IkB induction in hepatocytes was induced by injection of the synthetic steroid mifepristone. Activation of NF-kB in hepatocytes was suppressed. This, however, did not have any effect on liver regeneration, and there was no measurable impact on hepatocyte proliferation, function or viability. The study concluded that the effects of TNF (through NF-kB) on liver regeneration are mediated not through hepatocytes but through non-parenchymal cells, most likely Kupffer cells [177]. It should also be noted that even though regeneration in TNFR1 mice is deficient, it is nonetheless not absent; liver weight is eventually restored. In addition, EGFR and cMet can also cause activation of NF-kB in hepatocyte cultures (although not as efficiently as TNF), suggesting that the associated growth factors (HGF, EGF, etc.) can function as alternate "rescue" activators for NF-kB in the absence of TNF [178]. The role of TNF in liver regeneration overall appears to be associated with "priming" the hepatocytes to respond to directly mitogenic growth factors.

One of the effects of TNF is induction of expression of IL6. Administration of IL6 in TNFR1 deficient mice restored some of the observed deficiencies in liver regeneration [34]. IL6 is associated with induction of the spectrum of gene expression changes collectively known as "acute phase response", in which hepatocytes massively produce a spectrum of proteins in response to systemic acute inflammation or infection [179, 180]. Mice with homozygous deletions in IL6 do not have any demonstrable phenotypic abnormalities; during liver regeneration, however, they have deficient activation of Stat-3, a member of the "Signal transducer and activator of transcription" family of transcription factors [26, 181]. Stat-3 can be activated by other growth factors and cytokines in addition to IL6. These include other members of the cytokine family utilizing the shared receptor gp130 (such as Oncostatin M, LIF, CNTF) [182, 183], ligands for receptor tyrosine kinases (EGF, PDGF, HGF, etc.) [27, 184–186], non-receptor tyrosine kinases such as src [186], etc. IL6 is the strongest induced of Stat-3 in hepatocytes [29]. Mice with homozygous deletion of Stat-3 are more sensitive to massive hepatocyte apoptosis induced by activation of the apoptotic Fas receptor [29]. An earlier study suggested that IL6 –/– mice had deficient liver response to PHx, associated with areas of hepatic necrosis [181]. A subsequent study however showed that the response was deficient in a very minor way (5% difference in hepatocyte labeling) and that regeneration was proceeding relatively normally in IL6 –/– mice [187]. Further studies utilized mice with conditional deletion of the IL6 shared receptor gp130 targeted to the liver. In these mice, regenerative response to PHx of CCL4 was normal, unless the mice were injected at the same time with endotoxin [30, 188]. The latter was associated with a deficient regenerative response to IL –/– mice but not to wild type mice. These findings taken all together make it difficult to assess the role of IL6 in liver regeneration in otherwise normal mice. IL6 must play a role in this process, as evidenced by detail comparison of gene expression patterns between IL6 –/– and wild type mice at the first stages of liver regeneration [189]. A spectrum of genes are not properly expressed in the IL6 –/– mice during liver regeneration. It does not appear, however, that the IL6 –/– dependent responses are crucial for liver regeneration, though they undoubtedly contribute to the proper orchestration of the sequence of the early events. IL6 is not a direct mitogen for hepatocytes in culture, but it is a direct mitogen for biliary epithelial cells [13] and it is highly expressed in spontaneously growing organoid hepatic cultures [15]. Double transgenic mice with over-expression of both IL6 and its soluble receptor targeted to the liver, have liver enlargement and focal areas of hepatocyte hyperplasia [190].

## 5.7 Norepinephrine

This neurotransmitter of the sympathetic nervous system is also produced by the medullary portion of the adrenal glands and exercises effects on many cell types, including hepatocytes, through the alpha-1 adrenergic receptor (A1R). Norepinephrine is not a direct mitogen for hepatocytes in primary culture. It does, however, synergize and increase the mitogenic effects of EGF and HGF, by heterologous regulation of the EGF and HGF receptors via the A1R [191]. In addition, norepinephrine partially inhibits the mitoinhibitory effects of TGFβ1 on hepatocytes in culture. When norepinephrine is added to hepatocyte cultures kept in balance between EGF and TGFβ1, under conditions that there is little DNA synthesis, norepinephrine enhances the mitogenic effect of EGF and decreases the mitoinhibitory effect of TGFβ1, inducing entry of hepatocytes into DNA synthesis [192]. In addition to the modulating effects of norepinephrine on other cytokines and growth factors, norepinephrine also stimulates their production. Addition of norepinephrine to fibroblast cultures induces a dramatic upregulation of HGF gene expression [101]. Norepinephrine also increases production of EGF by the duodenal glands of Brunner, which can feed EGF directly to the liver via the portal circulation [60, 193]. Both of these effects may be of significance in liver regeneration, because norepinephrine and epinephrine levels in the plasma rise sharply within 30 min after PHx [131]. Norepinephrine thus may enhance the plasma and liver levels of mitogens, enhance their effects on hepatocytes and decrease the mitoinhibitory effects of TGFβ1. The above evidence for a potential role of norepinephrine is circumstantial. The case is strengthened however by the fact that blockade of the A1R by the specific inhibitor prazosin blocks DNA synthesis in hepatocytes after PHx in rats for the first 72 h [131]. A very interesting set of recent findings also suggests that norepinephrine exercises effects on stellate cells [194, 195]. Cultures of stellate cells produce norepinephrine and respond to it with synthesis of extracellular matrix proteins. Stellate cells from mice deficient in production of norepinephrine grow poorly in culture and their growth is rescued by addition of norepinephrine. Thus, the effects of norepinephrine appear global, affecting many aspects related to the ligands and receptors that trigger hepatocyte DNA synthesis.

## 5.8 Signaling System of Notch and Jagged

Notch as well as its binding ligands Jagged and Delta are all transmembrane proteins. Expression of both Notch and its ligands in adjacent cells facilitates contact through the plasma membrane and results in activation of Notch by specific proteases [196]. Activation of Notch leads to cleavage of the intracellular C-terminal Notch domain which migrates to the nucleus and acts as a coactivator transcription factor inducing gene expression of many genes related to cell cycle, including c-Myc and cyclin D1. Activation of Notch has effects on many cell types. Typically activation of Notch occurs in many terminally differentiated cell types prior to entry into G1. Malformations of the intrahepatic bile ducts seen in Alagille's syndrome are associated with mutations in Jagged-1 [197]. Recent studies [198] have shown that there is enhanced expression of Notch in hepatocytes after PHx. Recombinant Jagged protein induces DNA synthesis in hepatocyte cultures. Silencing RNA against Notch or Jagged causes a decrease in hepatocyte DNA synthesis during liver regeneration. Activated Notch (Notch intracellular domain) moves into hepatocyte nuclei within 15 min after PHx. The above findings suggest that the Notch/Jagged signaling system is also rapidly activated after PHx and contributes to the signaling events leading hepatocytes to proliferation.

## 5.9 Transforming Growth Factor β1 (TGFβ1)

This is a member of a subfamily of three members (TGFβ1–3). The broader family includes other cytokines and polypeptides such as activin, inhibin and bone morphogenetic proteins [199]. In addition to commonly shared amino acid sequence domains and overall structure, the receptors to these proteins also share common structure elements. The active form of TGFβ1 is produced after removal of another protein attached to TGFβ1 by disulphide bonds [200]. This is catalyzed, among other factors, by plasmin (generated by activation of plasminogen by urokinase, see above). The TGFb receptor is assembled by interaction between three plasma membrane proteins (TGFbR1–3). Binding of TGFβ1 to TGFbR2 induces serine phosphorylation of TGFbR1. This activates a serine-threonine kinase function which phosphorylates downstream targets of the Smad family of proteins [201]. TGFβ1 (and the other two members of the family) also share effects on responding cells. Mesenchymal cells typically respond to TGBFβ1 by increasing production of proteins of the extracellular matrix. TGFβ1 inhibits proliferation of most epithelial cells, including hepatocytes and counteracts effects of mitogens such HGF and EGF in a dose dependent manner [202, 203]. Norepinephrine decreases the mitoinhibitory effects of TGFβ1 (see above). TGFβ1 in the liver is synthesized by stellate cells [204, 205].

In order to assess properly the role of TGFβ1, the following aspects of its function need to be considered:

1. TGFβ1 is present in the extracellular matrix of most cells in its latent form, bound to the plasma membrane protein decorin [53].
2. Injection of an adenoviral construct carrying a dominant negative TGFβ1 receptor to the normal, unoperated, rat liver induced proliferation of hepatocytes [206].
3. TGFβ1 increases rapidly in the plasma after PHx, in the same pattern as HGF, at the same time as plasmin increases in the hepatic parenchyma, following increase of urokinase activity [3].

4. Plasma circulating active TGFβ1 is rapidly inactivated by binding to the plasma protein alpha-2-macroglobulin [207].
5. Immunoreactive TGFβ1 can be demonstrated on hepatocytes. During liver regeneration, the immunoreactivity is progressively lost from the periportal hepatocytes. The change of hepatocytes from TGF1 positive to TGFβ1 negative progresses as a wave towards the pericentral area of the lobule. Hepatocytes in mitosis can be seen behind the positive/negative immunoreactive interface, as it progresses from periportal to pericentral areas [208].
6. TGFβ1 mRNA increases in the liver starting at 3 h and peaking from 24 to 72 h after PHx [209].
7. DNA synthesis of normal hepatocytes in culture is inhibited by TGFβ1 in a dose dependent manner [202].
8. Hepatocytes from regenerating liver are resistant to the effects of TGFβ1 [192]. TGFβ1 receptor protein levels on hepatocytes in vivo decrease during liver regeneration [210].
9. In addition to its mitoinhibitory effects on hepatocytes, TGFβ1 also has strong motogenic effects. TGFβ1 alone stimulates hepatocyte migration across barriers. Its effects are additive to those of EGF. TGFβ1 and EGF added together have the same motogenic effect as HGF [211]. TGFβ1 and HGF induce an epithelial to mesenchymal transformation on mouse hepatocytes [212].
10. Finally, transgenic mice with targeted overexpression of TGFβ1 in hepatocytes have normal liver regeneration [213] but have increased fibrosis [214].

Many of the findings of 1–9 above may appear self contradictory and impossible to reconcile. Why does TGFβ1, a mitoinhibitor for hepatocytes, increase in the plasma after PHx and in the liver during regeneration? A scenario attempting to reconcile all the above findings is as follows.

TGFβ1 and HGF and EGF coexist in the extracellular matrix of the normal liver, bound to different moieties. Though they are bound, they must exercise some effect (perhaps by a slow rate of release and activation) that keeps hepatocytes of the normal liver in a resting, non-proliferating state, through a balance between mitogens (HGF, EGF) and mitoinhibitors (point 2, above). Following PHx, increase of urokinase and plasmin in the liver causes local release and activation of HGF and TGFβ1 and both activities increase in the plasma. Whereas active HGF is now available for mitogenic stimulation of hepatocytes, TGFβ1 is inactivated by alpha-2 macroglobulin and its mitoinhibitory effect is diminished (points 3 and 4 above). These events advance slowly through the hepatic parenchyma, during regeneration (point 5, above). The "new balance" between mitogens (HGF and EGF) and mitoinhibitors (TGFβ1) stimulates hepatocytes into proliferation. Other cytokines such as TNF and IL6 also rise and assist in progression of intracellular signaling cascades, mediated by activation of Stat-3 and NF-kB. Newly synthesized TGFβ1 is not mitoinhibitory to hepatocytes, perhaps by a combination of the effects of norepinephrine and the decrease in the levels of TGFβ1 receptor proteins (points 6, 7, 8 and 10 above).

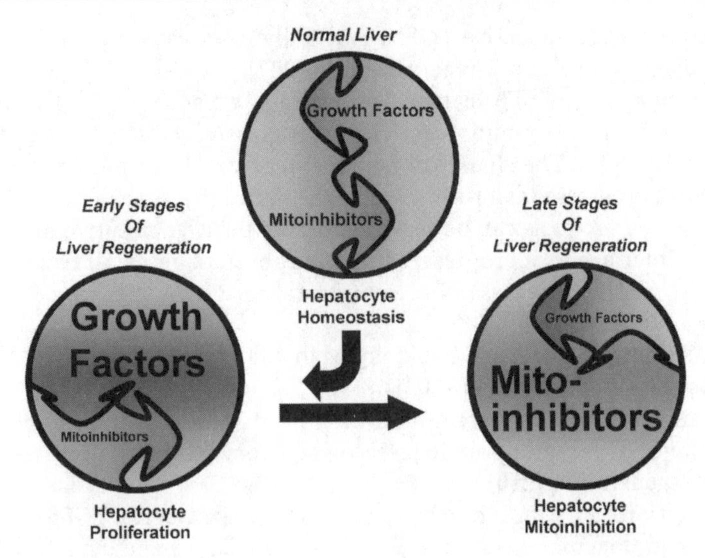


**Fig. 3** Cartoon depiction of the balancing factors leading hepatocytes to proliferation and return to quiescence

The effects of newly synthesized TGFβ1 during liver regeneration are probably related to its motogenic effects on hepatocytes and stimulation of formation of new extracellular matrix during and after liver regeneration by stellate cells (point 9 above). Several investigators have suggested that TGFβ1 may be involved in termination of liver regeneration, after liver mass has been restored. This remains an open possibility, even though this evidence is weakened by the fact that transgenic mice with overexpression of TGFβ1 in the liver have normal liver regeneration.

Overall, the state of hepatocytes (quiescent vs proliferative) appears to depend on rapid shifts in balance between signals leading to proliferation and signals leading to mitoinhibition (Fig. 3).

The role of the other members of TGFβ1 (TGFb2 and 3) is not clear, though they are expressed in liver and change during regeneration [215]. Of interest, similar effects to those of TGFβ1 have been described for proteins of the members of *activin* and *inhibin* [206, 216, 217]. Injection of follistatin (a protein that binds activin) to normal rats induces hepatocyte DNA synthesis and it accelerates hepatocyte DNA synthesis during regeneration [217, 218]. Injection into the liver (through the portal vein) of dominant negative activin receptor adenoviral constructs also causes hepatocyte DNA synthesis, suggesting a role for activin in normal liver similar to that of TGFβ1 that may keep hepatocytes of normal liver in a quiescent state [206, 217]. Chronic injection of activin leads to massive hepatocyte apoptosis [218, 219]. Overexpression of activin in the liver of transgenic mice has inhibitory effects but does not abolish regeneration [220]. Despite these effects, genetically modified mice with deficiencies in hepatic activin and inhibin do not have any hepatic abnormalities [221, 222].

## 6
## Termination of Liver Regeneration

While most of the research has concentrated on signaling pathways initiating liver regeneration, the precise mechanisms terminating the response are less understood. All evidence from work with animals points out that liver mass adjustment is precisely determined and that some degree of apoptosis at the end of the regenerative process may play a role [187]. When livers are transplanted from small dogs to big dogs, they grow up in size to that of the liver of a larger animal. The same phenomenon took place when livers from baboons were transplanted to immunosuppressed humans [223]. The reverse also occurs. Livers from large dogs transplanted into smaller animals adjust to a smaller size with a wave of controlled apoptosis. Several candidate processes and signals have emerged as potential contributors to the process of termination of regeneration. TGFβ1 (see above) does not have mitoinhibitory effects on hepatocytes during regeneration but may do so at later stages. Deposition of new matrix (in part stimulated by members of the TGFβ family) may also stop hepatocyte proliferation. Studies with hepatocyte cultures have demonstrated that extracellular matrix is always inhibitory to proliferation of hepatocyte monolayers [56, 224, 225]. It is not clear whether the full development of new sinusoid capillaries, (leading to higher oxygen levels, suppression of HIF1a, suppression of VEGF, suppression of HGF production by endothelial cells, decreased hepatocyte proliferation, etc.) might also play a role. The mechanisms controlling the proportionality between liver weight and body weight are also poorly understood. If a balance between mitogens (HGF, EGF) and mitoinhibitors (TGFβ1, activin) controls hepatocyte proliferation, then it is possible that a signaling pathway adjusting liver weight could operate if the mitogens bound to the hepatic biomatrix were derived from sources outside the liver (such as lung (for HGF) and duodenum (for EGF)). More mitogens per hepatocyte would be deposited to the matrix when a liver from a small animal were to be implanted into a large one, under that scenario, leading to imbalance between mitogens and mitoinhibitors and hepatocyte proliferation, new matrix synthesis, etc. until the balance was restored. The above, however, is purely speculative at this time and experimental work needs to be done to sort out several alternatives. It should also be noted that corticosteroids also suppress liver regeneration [226, 227]. This may occur by inhibitory effects on several mitogenic pathways [15]. The importance of this relative to liver regeneration is not clear.

## 7
## Alternate Pathways to Liver Regeneration

Several studies have shown that when hepatocyte proliferation is suppressed after PHx, then a new population of cells emerges from the periportal regions which eventually become hepatocytes and restore the hepatic mass. These cells have come under many names, with the one most prevalently used being that of "oval" cells (from the shape of the cells and its nucleus). The most commonly applied protocol for generation of oval cells is that of suppressing liver regeneration after partial hepatectomy by the compound N-acetylaminofluorene (AAF) [159, 228–231]. The latter is a genotoxic compound and it may be suppressing hepatocyte proliferation by activating pathways dependent on cell cycle regulator protein p53 [232, 233]. Oval cells emerge as small cells from around the portal triads and infiltrate the lobular areas. Pulse chase studies with tritiated thymidine have clearly shown that the oval cells become hepatocytes [229]. There is much evidence that the oval cells derive from cells associated with the biliary epithelium. The evidence is as follows:

1. Oval cells express many genes of both hepatocytes and biliary epithelial cells [234].
2. Hepatocyte associated transcription factors appear in biliary epithelial cells of portal ducts after PHx+AAF [235].
3. Lethal toxic injury to the biliary epithelium with DAPM prior to PHx+AAF prevents the emergence of oval cells and delays restoration of liver weight [4, 236].

Biliary epithelium, easily seen in the portal triads, does however extend beyond the portal area and comes into direct contact with hepatocytes in the form of the canals of Hering [237]. Bile is formed by the hepatocytes and it is excreted into the bile canaliculi (small canals between the plasma membrane of apposed hepatocytes). The canals of Hering are the first system of draining canals lined by biliary epithelial cells and they receive the bile directly from the bile canaliculi of the hepatocytes. It is highly likely that both the canals of Hering and the small portal ductules are precursors to oval cells. The above studies demonstrate that when hepatocyte proliferation is arrested, then *biliary epithelial cells become facultative stem cells for hepatocytes* and rescue liver regeneration. This phenomenon is of clinical relevance in human disease syndromes such as fulminant hepatitis, in which hepatocytes cannot regenerate fast enough to replenish the lost cells and the biliary compartment generates "ductular hepatocytes" (the human equivalent of oval cells) that may rescue the liver.

Recent evidence suggests that the opposite trend may also be true and that hepatocytes may be able to become biliary epithelial cells [238]. This was shown in organoid cultures derived from chimeric livers in which many hepatocytes were expressing specific markers unique to the donor strain (for details about the protocol leading to development of chimeric livers, see [239, 240]). Recent studies also show that this phenomenon occurs in the whole animal in protocols leading to bile duct injury and repair (Michalopoulos GK and Bowen WC: unpublished observations). These studies show that under selective conditions *hepatocytes may become facultative stem cells for the biliary epithelium*, rescuing the biliary compartment in situations where the biliary cells cannot act to repair biliary injury.

These findings suggest that, under normal conditions, there is no standing population of liver stem cells, analogous to the basal layer of the epidermis, the crypt cells of the small intestine or the hematopoietic stem cells. Studies trying to identify stem cells in the liver for many years have focused on transitional cell populations, derived not from pre-existing stem cells but from the adult epithelial cells of the liver (hepatocytes and biliary) undergoing transitions from one phenotype to the other.

There are few circumstances, primarily experimental, in which all cells of the liver have a compromised capacity to proliferate. This for example may occur after high levels of systemic radiation, or after administering agents that cross-link cell DNA, such as pyrrolizidine alkaloids (retrorsine, monocrotalline, etc.). Even after localized liver radiation with 1500 Rad, liver weight is restored after PHx. This, however, is not accomplished through hepatocyte proliferation. Instead of proliferating, hepatocytes increase their size and liver mass is restored by cellular hypertrophy [3]. Currently, none of the genetic deletions or manipulations used so far in experimental studies of liver regeneration is capable of abrogating the final endpoint, which is restoration of the total mass of the liver at the end of regeneration.

## 8 Hepatocyte Repopulation from Bone Marrow Hematopoietic Precursors

Bone marrow hematopoietic stem cells tagged with specific traceable markers were injected in several experimental models of liver injury and this resulted in the emergence of hepatocytes bearing the specific stem cell marker. This finding was most prominent with mice bearing genetic abnormalities causing liver failure. Mice with homozygous deletions of FAH were the most striking demonstration of this approach [241, 242]. Similar findings were also reported in other systems including heart, skeletal muscle, central nervous system, etc. The analysis of these studies however had to be re-evaluated when it was recently shown that all the apparent cell progeny of the bone marrow stem cells was actually the result of cell fusion between the injected bone marrow cells and the cells of the target tissues [243]. In the case of liver, careful karyotypic analysis of the resultant hepatocytes demonstrated that they were aneuploid [244–248]. The fusion between FAH –/– hepatocytes and FAH +/+ hematopoietic stem cells resulted in fusion products with hepatocyte phenotype that were FAH positive and thus could replicate and re-colonize the liver. The physiologic importance of these findings is not clear. It is widely believed that there are circulating hematopoietic stem cells but it not clear whether they do fuse with other normal cells in peripheral tissues under normal circumstances [246].

## 9 Hepatic Enlargement Induced by Hormones and Xenobiotics

Several xenobiotics induce enlargement of the liver to variable proportions, from barely noticeable to almost doubling of the size. Prototypes of this phenomenon are barbiturates [249] and compounds collectively known as peroxisome proliferators [250]. Similar effects are seen with thyroxine [251] and estradiol [252]. All of these compounds are ligands for nuclear hormone receptors such as CAR and PXR (for phenobarbital), members of the peroxisome proliferator associated receptors (PPAR alpha, beta, gamma and delta), the estrogen and thyroxine receptors, etc. Liver enlargement is mediated by a combination of cellular proliferation and hypetrophy [249]. Some of these compounds function as liver tumor promoters (e.g. phenobarbital, [253–256] whereas others are complete carcinogens by themselves (e.g. peroxisome proliferators). The mechanisms mediating these responses are not clear and it appears that the factors and signaling transduction pathways operating in the standard liver regeneration protocols are not operative in these phenomena [257, 258]. Liver mass is restored to normal by apoptosis after discontinuation of the exogenous compound and this is mediated via controlled hepatocyte apoptosis. The reader is referred to several reviews written for these xenobiotics for more in depth discussion of this phenomenon [259].

## 10 Conclusions

Regeneration of liver after partial hepatectomy or chemical injury is a model for understanding the complexity of the signaling pathways controlling tissue growth in adult organisms. The complexity and the redundancy of the pathways, both cellular and molecular, ensure that restoration of the hepatic mass will occur even if some of the standard components of these pathways become compromised. A combination of in vivo models, genetic deletions and complex cell culture continues to reveal more secrets of the complex process of hepatic tissue assembly. Despite the accumulated knowledge regarding this process, unrecoverable liver failure continues to remain a common cause of death in humans. It is hoped that in the years to come, therapeutic models will emerge from the knowledge gained from the experimental studies,

## References

1. Miller EC, Miller JA, Boberg EW, Delclos KB, Lai CC, Fennell TR, Wiseman RW, Liem A (1985) Carcinog Compr Surv 10:93
2. Potter WZ, Thorgeirsson SS, Jollow DJ, Mitchell JR (1974) Pharmacology 12:129
3. Michalopoulos GK, DeFrances MC (1997) Science 276:60

4. Petersen BE, Zajac VF, Michalopoulos GK (1998) Hepatology 27:1030
5. Higgins GM, Anderson RM (1931) Arch Pathol 12:186
6. Friedman SL, Roll FJ, Boyles J, Bissell DM (1985) Proc Natl Acad Sci USA 82:8681
7. Rabes HM (1977) Ciba Found Symp 31
8. Grisham J (1962) Cancer Res 22:842
9. Schirmacher P, Geerts A, Jung W, Pietrangelo A, Rogler CE, Dienes HP (1993) Exs 65:285
10. LeCouter J, Moritz DR, Li B, Phillips GL, Liang XH, Gerber HP, Hillan KJ, Ferrara N (2003) Science 299:890
11. Tomiya T, Ogata I, Yamaoka M, Yanase M, Inoue Y, Fujiwara K (2000) Am J Pathol 157:1693
12. Streetz KL, Wustefeld T, Klein C, Manns MP, Trautwein C (2001) Cell Mol Biol (Noisy-le-grand) 47:661
13. Matsumoto K, Fujii H, Michalopoulos G, Fung JJ, Demetris AJ (1994) Hepatology 20:376
14. Michalopoulos GK, Bowen WC, Mule K, Stolz DB (2001) Am J Pathol 159:1877
15. Michalopoulos GK, Bowen WC, Mule K, Luo J (2003) Gene Expr 11:55
16. Mars WM, Liu ML, Kitson RP, Goldfarb RH, Gabauer MK, Michalopoulos GK (1995) Hepatology 21:1695
17. Minuk GY, Kren BT, Xu R, Zhang X, Burczynski F, Mulrooney NP, Fan G, Gong Y, Steer CJ (1997) Hepatology 25:1123
18. Minuk GY (1993) Dig Dis 11:45
19. Stolz DB, Mars WM, Petersen BE, Kim TH, Michalopoulos GK (1999) Cancer Res 59:3954
20. Jo M, Stolz DB, Esplen JE, Dorko K, Michalopoulos GK, Strom SC (2000) J Biol Chem 275:8806
21. Haber B, Naji L, Cressman D, Taub R (1995) Hepatology 22:906
22. Haber BA, Mohn KL, Diamond RH, Taub R (1993) J Clin Invest 91:1319
23. Mohn KL, Laz TM, Hsu JC, Melby AE, Bravo R, Taub R (1991) Mol Cell Biol 11:381
24. Lee J, Greenbaum L, Haber BA, Nagle D, Lee V, Miles V, Mohn KL, Bucan M, Taub R (1994) Hepatology 19:656
25. Leu JI, Crissey MA, Craig LE, Taub R (2003) Mol Cell Biol 23:1251
26. Cressman DE, Diamond RH, Taub R (1995) Hepatology 21:1443
27. Runge DM, Runge D, Foth H, Strom SC, Michalopoulos GK (1999) Biochem Biophys Res Commun 265:376
28. Runge D, Runge DM, Drenning SD, Bowen WC Jr, Grandis JR, Michalopoulos GK (1998) Biochem Biophys Res Commun 250:762
29. Taub R (2003) J Clin Invest 112:978
30. Streetz KL, Wustefeld T, Klein C, Kallen KJ, Tronche F, Betz UA, Schutz G, Manns MP, Muller W, Trautwein C (2003) Gastroenterology 125:532
31. Streetz KL, Tacke F, Leifeld L, Wustefeld T, Graw A, Klein C, Kamino K, Spengler U, Kreipe H, Kubicka S, Muller W, Manns MP, Trautwein C (2003) Hepatology 38:218
32. Karin M, Yamamoto Y, Wang QM (2004) Nat Rev Drug Discov 3:17
33. Yamada Y, Webber EM, Kirillova I, Peschon JJ, Fausto N (1998) Hepatology 28:959
34. Yamada Y, Kirillova I, Peschon JJ, Fausto N (1997) Proc Natl Acad Sci USA 94:1441
35. Monga SP, Pediaditakis P, Mule K, Stolz DB, Michalopoulos GK (2001) Hepatology 33:1098
36. Monga SP, Mars WM, Pediaditakis P, Bell A, Mule K, Bowen WC, Wang X, Zarnegar R, Michalopoulos GK (2002) Cancer Res 62:2064
37. Hsu HC, Jeng YM, Mao TL, Chu JS, Lai PL, Peng SY (2000) Am J Pathol 157:763
38. Mao TL, Chu JS, Jeng YM, Lai PL, Hsu HC (2001) J Pathol 193:95
39. de La Coste A, Romagnolo B, Billuart P, Renard CA, Buendia MA, Soubrane O, Fabre M, Chelly J, Beldjord C, Kahn A, Perret C (1998) Proc Natl Acad Sci USA 95:8847

40. Braun L, Mikumo R, Fausto N (1989) Cancer Res 49:1554
41. Delahunty TJ, Rubinstein D (1970) J Lipid Res 11:536
42. Yee AG, Revel JP (1978) J Cell Biol 78:554
43. Andreasen PA, Egelund R, Petersen HH (2000) Cell Mol Life Sci 57:25
44. Kim TH, Mars WM, Stolz DB, Petersen BE, Michalopoulos GK (1997) Hepatology 26:896
45. Kim TH, Mars WM, Stolz DB, Michalopoulos GK (2000) Hepatology 31:75
46. Preaux AM, Mallat A, Nhieu JT, D'Ortho MP, Hembry RM, Mavier P (1999) Hepatology 30:944
47. Roselli HT, Su M, Washington K, Kerins DM, Vaughan DE, Russell WE (1998) Am J Physiol 275:G1472
48. Bezerra JA, Bugge TH, Melin-Aldana H, Sabla G, Kombrinck KW, Witte DP, Degen JL (1999) Proc Natl Acad Sci USA 96:15143
49. Visse R, Nagase H (2003) Circ Res 92:827
50. Schatzmann F, Marlow R, Streuli CH (2003) J Mammary Gland Biol Neoplasia 8:395
51. Swindle CS, Tran KT, Johnson TD, Banerjee P, Mayes AM, Griffith L, Wells A (2001) J Cell Biol 154:459
52. Filla MS, Dam P, Rapraeger AC (1998) J Cell Physiol 174:310
53. Kresse H, Schonherr E (2001) J Cell Physiol 189:266
54. Kleinman HK, McGarvey ML, Liotta LA, Robey PG, Tryggvason K, Martin GR (1982) Biochemistry 21:6188
55. Michalopoulos GK, Bowen WC, Zajac VF, Beer-Stolz D, Watkins S, Kostrubsky V, Strom SC (1999) Hepatology 29:90
56. Rana B, Mischoulon D, Xie Y, Bucher NL, Farmer SR (1994) Mol Cell Biol 14:5858
57. Liu ML, Mars WM, Zarnegar R, Michalopoulos GK (1994) Hepatology 19:1521
58. Carpenter G, Cohen S (1990) J Biol Chem 265:7709
59. Gill GN (1990) Mol Reprod Dev 27:46
60. Olsen PS, Poulsen SS, Kirkegaard P (1985) Gut 26:920
61. McGowan JA, Strain AJ, Bucher NL (1981) J Cell Physiol 108:353
62. St Hilaire RJ, Jones AL (1982) Hepatology 2:601
63. Bucher NL, Patel U, Cohen S (1977) Adv Enzyme Regul 16:205
64. Bucher NL, Patel U, Cohen S (1977) Ciba Found Symp 55:95
65. Jones DE Jr, Tran-Patterson R, Cui DM, Davin D, Estell KP, Miller DM (1995) Am J Physiol 268:G872
66. Lambotte L, Saliez A, Triest S, Maiter D, Baranski A, Barker A, Li B (1997) Hepatology 25:607
67. Luetteke NC, Lee DC (1990) Semin Cancer Biol 1:265
68. Lee DC, Sunnarborg SW, Hinkle CL, Myers TJ, Stevenson MY, Russell WE, Castner BJ, Gerhart MJ, Paxton RJ, Black RA, Chang A, Jackson LF (2003) Ann NY Acad Sci 995:22
69. Wong ST, Winchell LF, McCune BK, Earp HS, Teixido J, Massague J, Herman B, Lee DC (1989) Cell 56:495
70. Lee DC, Fenton SE, Berkowitz EA, Hissong MA (1995) Pharmacol Rev 47:51
71. Reddy CC, Wells A, Lauffenburger DA (1996) J Cell Physiol 166:512
72. Reddy CC, Wells A, Lauffenburger DA (1998) Med Biol Eng Comput 36:499
73. Mead JE, Fausto N (1989) Proc Natl Acad Sci USA 86:1558
74. Webber EM, Bruix J, Pierce RH, Fausto N (1998) Hepatology 28:1226
75. Webber EM, Wu JC, Wang L, Merlino G, Fausto N (1994) Am J Pathol 145:398
76. Lee GH, Merlino G, Fausto N (1992) Cancer Res 52:5162
77. Russell WE, Dempsey PJ, Sitaric S, Peck AJ, Coffey RJ Jr (1993) Endocrinology 133:1731
78. Luetteke NC, Qiu TH, Peiffer RL, Oliver P, Smithies O, Lee DC (1993) Cell 73:263

79. Russell WE, Kaufmann WK, Sitaric S, Luetteke NC, Lee DC (1996) Mol Carcinog 15:183
80. Kiso S, Kawata S, Tamura S, Umeki S, Ito N, Tsushima H, Yamada A, Miyagawa J, Higashiyama S, Taniguchi N, Matsuzawa Y (1999) Biochem Biophys Res Commun 259:683
81. Ito N, Kawata S, Tamura S, Kiso S, Tsushima H, Damm D, Abraham JA, Higashiyama S, Taniguchi N, Matsuzawa Y (1994) Biochem Biophys Res Commun 198:25
82. Kiso S, Kawata S, Tamura S, Miyagawa J, Ito N, Tsushima H, Yamada A, Umeki S, Higashiyama S, Taniguchi N, Matsuzawa Y (1999) J Gastroenterol Hepatol 14:1203
83. Kiso S, Kawata S, Tamura S, Ito N, Tsushima H, Yamada A, Higashiyama S, Taniguchi N, Matsuzawa Y (1996) Biochem Biophys Res Commun 220:285
84. Komurasaki T, Toyoda H, Uchida D, Nemoto N (2002) Growth Factors 20:61
85. Sibilia M, Steinbach JP, Stingl L, Aguzzi A, Wagner EF (1998) Embo J 17:719
86. Carver RS, Mathew PM, Russell WE (1997) Endocrinology 138:5195
87. Carver RS, Sliwkowski MX, Sitaric S, Russell WE (1996) J Biol Chem 271:13491
88. Carver RS, Stevenson MC, Scheving LA, Russell WE (2002) Gastroenterology 123: 2017
89. Zarnegar R, Michalopoulos G (1989) Cancer Res 49:3314
90. Michalopoulos G, Houck KA, Dolan ML, Leutteke NC (1984) Cancer Res 44:4414
91. Nakamura T, Nishizawa T, Hagiya M, Seki T, Shimonishi M, Sugimura A, Tashiro K, Shimizu S (1989) Nature 342:440
92. Leong GF, Grisham JW, Hole BV, Albright ML (1964) Cancer Res 24:1496
93. Jirtle RL, Michalopoulos G (1982) Cancer Res 42:3000
94. Moolten FL, Bucher NL (1967) Science 158:272
95. Naldini L, Vigna E, Narsimhan RP, Gaudino G, Zarnegar R, Michalopoulos GK, Comoglio PM (1991) Oncogene 6:501
96. Bottaro DP, Rubin JS, Faletto DL, Chan AM, Kmiecik TE, Vande Woude GF, Aaronson SA (1991) Science 251:802
97. Gherardi E, Sharpe M, Lane K, Sirulnik A, Stoker M (1993) Symp Soc Exp Biol 47:163
98. Matsumoto K, Nakamura T (1996) J Biochem (Tokyo) 119:591
99. Galimi F, Brizzi MF, Comoglio PM (1993) Stem Cells 11 Suppl 2:22
100. Vigna E, Naldini L, Tamagnone L, Longati P, Bardelli A, Maina F, Ponzetto C, Comoglio PM (1994) Cell Mol Biol (Noisy-le-grand) 40:597
101. Broten J, Michalopoulos G, Petersen B, Cruise J (1999) Biochem Biophys Res Commun 262:76
102. Achim CL, Katyal S, Wiley CA, Shiratori M, Wang G, Oshika E, Petersen BE, Li JM, Michalopoulos GK (1997) Brain Res Dev Brain Res 102:299
103. Maggiora P, Gambarotta G, Olivero M, Giordano S, Di Renzo MF, Comoglio PM (1997) J Cell Physiol 173:183
104. Nakamura T (1994) Princess Takamatsu Symp 24:195
105. Pollack AL, Apodaca G, Mostov KE (2004) Am J Physiol Cell Physiol 286:C482
106. Han S, Stuart LA, Degen SJ (1991) Biochemistry 30:9768
107. Bezerra JA, Witte DP, Aronow BJ, Degen SJ (1993) Hepatology 18:394
108. Leonard EJ (1997) Ciba Found Symp 212:183
109. Masumoto A, Yamamoto N (1991) Biochem Biophys Res Commun 174:90
110. Liu ML, Mars WM, Zarnegar R, Michalopoulos GK (1994) Am J Pathol 144:129
111. Mars WM, Zarnegar R, Michalopoulos GK (1993) Am J Pathol 143:949
112. Shimizu M, Hara A, Okuno M, Matsuno H, Okada K, Ueshima S, Matsuo O, Niwa M, Akita K, Yamada Y, Yoshimi N, Uematsu T, Kojima S, Friedman SL, Moriwaki H, Mori H (2001) Hepatology 33:569
113. Powell EM, Mars WM, Levitt P (2001) Neuron 30:79

114. Matsubara Y, Ichinose M, Yahagi N, Tsukada S, Oka M, Miki K, Kimura S, Omata M, Shiokawa K, Kitamura N, Kaneko Y, Fukamachi H (1998) Biochem Biophys Res Commun 253:477
115. Cioce V, Csaky KG, Chan AM, Bottaro DP, Taylor WG, Jensen R, Aaronson SA, Rubin JS (1996) J Biol Chem 271:13110
116. Pediaditakis P, Monga SP, Mars WM, Michalopoulos GK (2002) J Biol Chem 277:14109
117. Kosai K, Matsumoto K, Nagata S, Tsujimoto Y, Nakamura T (1998) Biochem Biophys Res Commun 244:683
118. Wang X, DeFrances MC, Dai Y, Pediaditakis P, Johnson C, Bell A, Michalopoulos GK, Zarnegar R (2002) Mol Cell 9:411
119. Wolf HK, Michalopoulos GK (1992) Hepatology 15:707
120. Shiota G, Okano J, Kawasaki H, Kawamoto T, Nakamura T (1995) Hepatology 21:106
121. Tsubouchi H, Hirono S, Gohda E, Nakayama H, Takahashi K, Sakiyama O, Kimoto M, Kawakami S, Miyoshi H, Kubozono O, et al. (1991) Dig Dis Sci 36:780
122. Tsubouchi H, Niitani Y, Hirono S, Nakayama H, Gohda E, Arakaki N, Sakiyama O, Takahashi K, Kimoto M, Kawakami S, et al. (1991) Hepatology 13:1
123. Tomiya T, Nagoshi S, Fujiwara K (1992) Hepatology 15:1
124. Lindroos PM, Zarnegar R, Michalopoulos GK (1991) Hepatology 13:743
125. Tomiya T, Tani M, Yamada S, Hayashi S, Umeda N, Fujiwara K (1992) Gastroenterology 103:1621
126. Kawaguchi Y, Harigai M, Fukasawa C, Hara M (1999) J Rheumatol 26:1012
127. Balkovetz DF, Lipschutz JH (1999) Int Rev Cytol 186:225
128. Sugimura K, Lee CC, Kim T, Goto T, Kasai S, Harimoto K, Yamagami S, Kishimoto T (1997) Nephron 75:7
129. Zarnegar R, DeFrances MC, Kost DP, Lindroos P, Michalopoulos GK (1991) Biochem Biophys Res Commun 177:559
130. Matsumoto K, Nakamura T (1993) Exs 65:225
131. Cruise JL, Knechtle SJ, Bollinger RR, Kuhn C, Michalopoulos G (1987) Hepatology 7:1189
132. Higashitsuji H, Arii S, Furutani M, Mise M, Monden K, Fujita S, Ishiguro S, Kitao T, Nakamura T, Nakayama H et al. (1995) J Surg Res 58:267
133. Yamada Y, Fausto N (1998) Am J Pathol 152:1577
134. Pediaditakis P, Lopez-Talavera JC, Petersen B, Monga SP, Michalopoulos GK (2001) Hepatology 34:688
135. Patijn GA, Lieber A, Schowalter DB, Schwall R, Kay MA (1998) Hepatology 28:707
136. Yang J, Chen S, Huang L, Michalopoulos GK, Liu Y (2001) Hepatology 33:848
137. Burr AW, Toole K, Chapman C, Hines JE, Burt AD (1998) J Pathol 185:298
138. Schmidt C, Bladt F, Goedecke S, Brinkmann V, Zschiesche W, Sharpe M, Gherardi E, Birchmeier C (1995) Nature 373:699
139. Brand-Saberi B, Muller TS, Wilting J, Christ B, Birchmeier C (1996) Dev Biol 179:303
140. Birchmeier C, Gherardi E (1998) Trends Cell Biol 8:404
141. Block GD, Locker J, Bowen WC, Petersen BE, Katyal S, Strom SC, Riley T, Howard TA, Michalopoulos GK (1996) J Cell Biol 132:1133
142. Schirmacher P, Geerts A, Pietrangelo A, Dienes HP, Rogler CE (1992) Hepatology 15:5
143. Scheving LA, Stevenson MC, Taylormoore JM, Traxler P, Russell WE (2002) Biochem Biophys Res Commun 290:197
144. Ueki T, Kaneda Y, Tsutsui H, Nakanishi K, Sawa Y, Morishita R, Matsumoto K, Nakamura T, Takahashi H, Okamoto E, Fujimoto J (1999) Nat Med 5:226
145. Matsuda Y, Matsumoto K, Yamada A, Ichida T, Asakura H, Komoriya Y, Nishiyama E, Nakamura T (1997) Hepatology 26:81

146. McKeehan WL, Wang F, Kan M (1998) Prog Nucleic Acid Res Mol Biol 59:135
147. McKeehan WL, Kan M (1994) Mol Reprod Dev 39:69
148. Houck KA, Zarnegar R, Muga SJ, Michalopoulos GK (1990) J Cell Physiol 143:129
149. Coumoul X, Deng CX (2003) Birth Defects Res Part C Embryo Today 69:286
150. Kan M, Huang JS, Mansson PE, Yasumitsu H, Carr B, McKeehan WL (1989) Proc Natl Acad Sci USA 86:7432
151. Yu C, Wang F, Jin C, Huang X, Miller DL, Basilico C, McKeehan WL (2003) Am J Pathol 163:1653
152. Yu C, Wang F, Jin C, Wu X, Chan WK, McKeehan WL (2002) Am J Pathol 161:2003
153. Jung J, Zheng M, Goldfarb M, Zaret KS (1999) Science 284:1998
154. Rossi JM, Dunn NR, Hogan BL, Zaret KS (2001) Genes Dev 15:1998
155. Strain AJ, McGuinness G, Rubin JS, Aaronson SA (1994) Exp Cell Res 210:253
156. Nguyen HQ, Danilenko DM, Bucay N, DeRose ML, Van GY, Thomason A, Simonet WS (1996) Oncogene 12:2109
157. Galli SJ, Zsebo KM, Geissler EN (1994) Adv Immunol 55:1
158. Ren X, Hogaboam C, Carpenter A, Colletti L (2003) J Clin Invest 112:1407
159. Fujio K, Evarts RP, Hu Z, Marsden ER, Thorgeirsson SS (1994) Lab Invest 70:511
160. Martinez-Hernandez A, Amenta PS (1995) Faseb J 9:1401
161. Ross MA, Sander CM, Kleeb TB, Watkins SC, Stolz DB (2001) Hepatology 34:1135
162. Ishikawa K, Mochida S, Mashiba S, Inao M, Matsui A, Ikeda H, Ohno A, Shibuya M, Fujiwara K (1999) Biochem Biophys Res Commun 254:587
163. Mochida S, Ishikawa K, Inao M, Shibuya M, Fujiwara K (1996) Biochem Biophys Res Commun 226:176
164. Kraizer Y, Mawasi N, Seagal J, Paizi M, Assy N, Spira G (2001) Biochem Biophys Res Commun 287:209
165. Sato T, El-Assal ON, Ono T, Yamanoi A, Dhar DK, Nagasue N (2001) J Hepatol 34:690
166. Camussi G, Montrucchio G, Lupia E, Soldi R, Comoglio PM, Bussolino F (1997) J Immunol 158:1302
167. Wack KE, Ross MA, Zegarra V, Sysko LR, Watkins SC, Stolz DB (2001) Hepatology 33:363
168. Drixler TA, Vogten MJ, Ritchie ED, van Vroonhoven TJ, Gebbink MF, Voest EE, Borel Rinkes IH (2002) Ann Surg 236:703
169. Malik R, Selden C, Hodgson H (2002) Semin Cell Dev Biol 13:425
170. Panesar N, Tolman K, Mazuski JE (1999) J Surg Res 85:251
171. Akerman P, Cote P, Yang SQ, McClain C, Nelson S, Bagby GJ, Diehl AM (1992) Am J Physiol 263:G579
172. Fausto N (2000) J Hepatol 32:19
173. Kirillova I, Chaisson M, Fausto N (1999) Cell Growth Differ 10:819
174. Bellas RE, FitzGerald MJ, Fausto N, Sonenshein GE (1997) Am J Pathol 151:891
175. Taub R (1996) Faseb J 10:413
176. Iimuro Y, Nishiura T, Hellerbrand C, Behrns KE, Schoonhoven R, Grisham JW, Brenner DA (1998) J Clin Invest 101:802
177. Chaisson ML, Brooling JT, Ladiges W, Tsai S, Fausto N (2002) J Clin Invest 110:193
178. Muller M, Morotti A, Ponzetto C (2002) Mol Cell Biol 22:1060
179. Gauldie J, Northemann W, Fey GH (1990) J Immunol 144:3804
180. Heinrich PC, Castell JV, Andus T (1990) Biochem J 265:621
181. Cressman DE, Greenbaum LE, DeAngelis RA, Ciliberto G, Furth EE, Poli V, Taub R (1996) Science 274:1379
182. Guillet C, Lelievre E, Plun-Favreau H, Froger J, Chabbert M, Hermann J, Benoit de Coignac A, Bonnefoy JY, Gascan H, Gauchat JF, Elson G (2002) Eur J Biochem 269:1932
183. Repovic P, Fears CY, Gladson CL, Benveniste EN (2003) Oncogene 22:8117

184. Maffe A, Comoglio PM (1998) Eur J Morphol 36 Suppl 74
185. Runge DM, Runge D, Dorko K, Pisarov LA, Leckel K, Kostrubsky VE, Thomas D, Strom SC, Michalopoulos GK (1999) J Hepatol 30:265
186. Boccaccio C, Ando M, Tamagnone L, Bardelli A, Michieli P, Battistini C, Comoglio PM (1998) Nature 391:285
187. Sakamoto T, Liu Z, Murase N, Ezure T, Yokomuro S, Poli V, Demetris AJ (1999) Hepatology 29:403
188. Wuestefeld T, Klein C, Streetz KL, Betz U, Lauber J, Buer J, Manns MP, Mueller W, Trautwein C (2002) J Biol Chem 30:30
189. Li W, Liang X, Leu JI, Kovalovich K, Ciliberto G, Taub R (2001) Hepatology 33:1377
190. Maione D, Di Carlo E, Li W, Musiani P, Modesti A, Peters M, Rose-John S, Della Rocca C, Tripodi M, Lazzaro D, Taub R, Savino R, Ciliberto G (1998) EMBO J 17:5588
191. Cruise JL, Houck KA, Michalopoulos GK (1985) Science 227:749
192. Houck KA, Michalopoulos GK (1989) J Cell Physiol 141:503
193. Skov Olsen P, Boesby S, Kirkegaard P, Therkelsen K, Almdal T, Poulsen SS, Nexo E (1988) Hepatology 8:992
194. Oben JA, Roskams T, Yang S, Lin H, Sinelli N, Torbenson M, Smedh U, Moran TH, Li Z, Huang J, Thomas SA, Diehl AM (2004) Gut 53:438
195. Oben JA, Yang S, Lin H, Ono M, Diehl AM (2003) Biochem Biophys Res Commun 302:685
196. Iso T, Hamamori Y, Kedes L (2003) Arterioscler Thromb Vasc Biol 23:543
197. McCright B, Lozier J, Gridley T (2002) Development 129:1075
198. Köhler C, Bell A, Bowen W, Monga S, Fleig W, Michalopoulos G (2004) Hepatology 39:1056
199. Mehra A, Wrana JL (2002) Biochem Cell Biol 80:605
200. Abe M, Oda N, Sato Y, Shibata K, Yamasaki M (2002) Endothelium 9:25
201. de Caestecker M (2004) Cytokine Growth Factor Rev 15:1
202. Carr BI, Hayashi I, Branum EL, Moses HL (1986) Cancer Res 46:2330
203. Houck KA, Cruise JL, Michalopoulos G (1988) J Cell Physiol 135:551
204. Ikeda H, Nagoshi S, Ohno A, Yanase M, Maekawa H, Fujiwara K (1998) Biochem Biophys Res Commun 250:769
205. Bissell DM (1998) J Gastroenterol 33:295
206. Ichikawa T, Zhang YQ, Kogure K, Hasegawa Y, Takagi H, Mori M, Kojima I (2001) Hepatology 34:918
207. Arandjelovic S, Freed TA, Gonias SL (2003) Biochemistry 42:6121
208. Jirtle RL, Carr BI, Scott CD (1991) J Biol Chem 266:22444
209. Fausto N, Mead JE, Braun L, Thompson NL, Panzica M, Goyette M, Bell GI, Shank PR (1986) Symp Fundam Cancer Res 39:69
210. Chari RS, Price DT, Sue SR, Meyers WC, Jirtle RL (1995) Am J Surg 169:126
211. Stolz DB, Michalopoulos GK (1994) J Cell Biochem 55:445
212. Petersen B, Yee CJ, Bowen W, Zarnegar R, Michalopoulos GK (1994) Cell Biol Toxicol 10:219
213. Sanderson N, Factor V, Nagy P, Kopp J, Kondaiah P, Wakefield L, Roberts AB, Sporn MB, Thorgeirsson SS (1995) Proc Natl Acad Sci USA 92:2572
214. Ueberham E, Low R, Ueberham U, Schonig K, Bujard H, Gebhardt R (2003) Hepatology 37:1067
215. Jakowlew SB, Mead JE, Danielpour D, Wu J, Roberts AB, Fausto N (1991) Cell Regul 2:535
216. Yasuda H, Mine T, Shibata H, Eto Y, Hasegawa Y, Takeuchi T, Asano S, Kojima I (1993) J Clin Invest 92:1491
217. Kogure K, Zhang YQ, Maeshima A, Suzuki K, Kuwano H, Kojima I (2000) Hepatology 31:916

218. Schwall RH, Robbins K, Jardieu P, Chang L, Lai C, Terrell TG (1993) Hepatology 18:347
219. Hully JR, Chang L, Schwall RH, Widmer HR, Terrell TG, Gillett NA (1994) Hepatology 20:854
220. Chabicovsky M, Herkner K, Rossmanith W (2003) Endocrinology 144:3497
221. Chang H, Lau AL, Matzuk MM (2001) Mol Cell Endocrinol 180:39
222. Pierson TM, Wang Y, DeMayo FJ, Matzuk MM, Tsai SY, Omalley BW (2000) Mol Endocrinol 14:1075
223. Manez R, Kelly RH, Marino IR, Murase N, Demetris AJ, Starzl TE (1994) Transplant Proc 26:1249
224. Michalopoulos GK, Bowen W, Nussler AK, Becich MJ, Howard TA (1993) J Cell Physiol 156:443
225. Kim TH, Bowen WC, Stolz DB, Runge D, Mars WM, Michalopoulos GK (1998) Exp Cell Res 244:93
226. Nagy P, Kiss A, Schnur J, Thorgeirsson SS (1998) Hepatology 28:423
227. Nagy P, Teramoto T, Factor VM, Sanchez A, Schnur J, Paku S, Thorgeirsson SS (2001) Hepatology 33:339
228. Evarts RP, Hu Z, Fujio K, Marsden ER, Thorgeirsson SS (1993) Cell Growth Differ 4:555
229. Evarts RP, Hu Z, Omori N, Omori M, Marsden ER, Thorgeirsson SS (1996) Carcinogenesis 17:2143
230. Thorgeirsson SS, Evarts RP, Bisgaard HC, Fujio K, Hu Z (1993) Proc Soc Exp Biol Med 204:253
231. Evarts RP, Nagy P, Nakatsukasa H, Marsden E, Thorgeirsson SS (1989) Cancer Res 49:1541
232. Lindeman B, Skarpen E, Oksvold MP, Huitfeldt HS (2000) Mol Carcinog 27:190
233. Trautwein C, Will M, Kubicka S, Rakemann T, Flemming P, Manns MP (1999) Oncogene 18:6443
234. Vandersteenhoven AM, Burchette J, Michalopoulos G (1990) Arch Pathol Lab Med 114:403
235. Bisgaard HC, Nagy P, Santoni-Rugiu E, Thorgeirsson SS (1996) Hepatology 23:62
236. Petersen BE, Zajac VF, Michalopoulos GK (1997) Am J Pathol 151:905
237. Crawford JM (2002) Semin Liver Dis 22:213
238. Michalopoulos GK, Bowen WC, Mule K, Lopez-Talavera JC, Mars W (2002) Hepatology 36:278
239. Laconi S, Pillai S, Porcu PP, Shafritz DA, Pani P, Laconi E (2001) Am J Pathol 158:771
240. Laconi E, Oren R, Mukhopadhyay DK, Hurston E, Laconi S, Pani P, Dabeva MD, Shafritz DA (1998) Am J Pathol 153:319
241. Grompe M, Al-Dhalimy M, Finegold M, Ou CN, Burlingame T, Kennaway NG, Soriano P (1993) Genes Dev 7:2298
242. Overturf K, Al-Dhalimy M, Ou CN, Finegold M, Grompe M (1997) Am J Pathol 151:1273
243. Alvarez-Dolado M, Pardal R, Garcia-Verdugo JM, Fike JR, Lee HO, Pfeffer K, Lois C, Morrison SJ, Alvarez-Buylla A (2003) Nature 425:968
244. Wang X, Willenbring H, Akkari Y, Torimaru Y, Foster M, Al-Dhalimy M, Lagasse E, Finegold M, Olson S, Grompe M (2003) Nature 422:897
245. Wang X, Foster M, Al-Dhalimy M, Lagasse E, Finegold M, Grompe M (2003) Proc Natl Acad Sci USA 100 Suppl 1:11881
246. Grompe M (2003) Semin Liver Dis 23:363
247. Vassilopoulos G, Russell DW (2003) Curr Opin Genet Dev 13:480
248. Vassilopoulos G, Wang PR, Russell DW (2003) Nature 422:901
249. Michalopoulos G (1991) Prog Clin Biol Res 369:227
250. Reddy JK, Rao MS (1992) IARC Sci Publ 116:225

251. Short J, Wedmore R, Kibert L, Zemel R (1980) Cytobios 28:165
252. Yager JD, Shi YE (1991) Prog Clin Biol Res 369:53
253. Jirtle RL, Meyer SA, Brockenbrough JS (1991) Prog Clin Biol Res 369:209
254. Ledda-Columbano GM, Coni P, Curto M, Giacomini L, Faa G, Sarma DS, Columbano A (1992) Carcinogenesis 13:379
255. Pitot HC, Goldsworthy T, Moran S, Sirica AE, Weeks J (1982) Carcinog Compr Surv 7:85
256. Dragan YP, Sargent L, Xu YD, Xu YH, Pitot HC (1993) Proc Soc Exp Biol Med 202:16
257. Ledda-Columbano GM, Curto M, Piga R, Zedda AI, Menegazzi M, Sartori C, Shinozuka H, Bluethmann H, Poli V, Ciliberto G, Columbano A (1998) Oncogene 17:1039
258. Menegazzi M, Carcereri-De Prati A, Suzuki H, Shinozuka H, Pibiri M, Piga R, Columbano A, Ledda-Columbano GM (1997) Hepatology 25:585
259. Hikita H, Vaughan J, Babcock K, Pitot HC (1999) Toxicol Sci 52:17


Received: April 2004

Adv Biochem Engin/Biotechnol (2005) 93: 135–159
DOI 10.1007/b99969


# Stem Cells in CNS and Cardiac Regeneration

David L. Stocum (✉)

Department of Biology and Indiana University Center for Regenerative Biology and Medicine, Indiana University-Purdue University Indianapolis, 402 North Blackford St., Indianapolis, IN 46202, USA
*dstocum@iupui.edu*



**Abstract** The central nervous system (CNS) and the heart muscle regenerate poorly after injury, yet evidence is mounting that both harbor cells capable of rebuilding neural and cardiac tissue. The reason for the poor regenerative response of CNS tissue and myocardium must therefore lie in the nature of the injury environment, which promotes fibrosis over regeneration. Strategies for regenerating these tissues thus rely on overcoming the fibrotic response by filling lesions with tissue-specific regeneration-competent cells that replace or rescue dying cells, or by activating endogenous regeneration-competent cells that do likewise. There has also been considerable excitement about the possibility of transplanting bone marrow cells into CNS or cardiac lesions to repair them, because bone marrow cells have been reported to be pluripotent. In this chapter, contemporary evidence for the existence of regeneration-competent cells in the CNS and heart is discussed, as well as attempts to use these cells and bone marrow cells to reconstitute new tissue.

**Keywords** Regeneration · Central Nervous System · Cardiac Muscle · Regenerative Medicine · Transdifferentiation

# 1 Introduction

Adult stem cells are defined as undifferentiated clonogenic cells that can self-renew while giving rise to differentiated progeny [1–3]. Such stem cells can be divided into a long-term (LT) subpopulation that can self-renew indefinitely, and a short-term (ST) subpopulation of committed progenitors derived from the LT subpopulation that self-renew only for a restricted interval [4]. The LT subpopulation maintains the stem cell population at a constant level while the ST subpopulation is a transit amplifying population that increases the number of terminally differentiated cell products [5]. Stem cells can be multipotent (e.g., hematopoietic and mesenchymal stem cells, intestinal epithelial cells), bipotent (e.g., liver) or unipotent (e.g., epidermis, acoustic sensory epithelium). They are usually small cells with a high nuclear/cytoplasmic ratio and are characterized molecularly by the expression of type-specific combinations of transcription factors and cell surface antigens.

All the cell types of the adult body, including adult stem cells, are derived from fewer than 100 embryonic stem cells (ESCs) of the pre-implantation blastocyst. Embryonic stem cells give rise to the three germ layers of the embryo – ectoderm, endoderm and mesoderm – from which all of the more than 200 differentiated cell types of the body emerge. ESC lines have been established in vitro from the early embryos of fish, birds, and a variety of mammals, including humans [6–10]. Human ESCs express markers characteristic of primate pluripotent cells: stage-specific embryonic antigens (SSEA-3 and 4, TRA-1–60 and 81), alkaline phosphatase, and high levels of telomerase [9]. ESCs are pluripotent, as demonstrated by the fact that they make contributions to all tissues when injected into an early blastocyst to form a chimeric embryo [11]. The pluripotency of mouse ESCs requires expression of the leukemia inhibitory factor (LIF)/STAT3-dependent transcription factors OCT4 [11, 12], SOX2 [13] and Fox D3 [14]. Recently, a LIF/STAT3-independent transcription factor, Nanog, has been discovered that is also essential for mouse ESC pluripotency and self-renewal [15, 16].

The function of stem cells in adults is regeneration, which maintains normal levels of differentiated cells lost by turnover in the uninjured animal and restores normal numbers of cells and tissue architecture lost to damage by injury or disease. Maintenance regeneration is highly visible in tissues that turn over rapidly (e.g., blood, epithelial tissues), but much less so in tissues that turn over slowly (muscle, bone, hippocampus), and most likely involves the asymmetric division of stem cells. The number of stem cells is elevated during regenerative responses to injury, suggesting that at least a fraction of them initially divides symmetrically, rather than asymmetrically, to produce only stem cell progeny [1].

While most adult mammalian tissue regeneration is accomplished by resident stem cells, a survey of regenerative phenomena in vertebrates makes clear that not all the cells involved in regeneration fit the classic definition of a stem cell. The mammalian liver uses a population of small oval stem cells located in the bile ductules to regenerate after severe chemical damage to hepatocytes. However, maintenance regeneration of the liver, which turns over during the period of a year [17] and regeneration after surgical tissue removal are achieved by the compensatory hyperplasia of all its cell types [18]. Compensatory hyperplasia of a liver cell involves the activation of its cell cycle program while maintaining all differentiated functions. Thus, the hepatocyte is a cell that can self-renew in the differentiated state. In anuran tadpoles and urodele larvae and adults, a large number of body parts, including limbs, tails, jaws, heart, intestine, neural retina and lens can regenerate by the partial or full dedifferentiation of differentiated cells to form a blastema of proliferating cells that subsequently redifferentiate or transdifferentiate into new tissue [19, 20]. These parts can regenerate repeatedly because each new injury re-induces dedifferentiation. Thus, cells derived by dedifferentiation can be said to be self-renewing in the sense that they can be re-created in successive rounds of regeneration. Mouse muscle has been reported to dedifferentiate in vitro into single cells with some of the properties of mesenchymal stem cells when treated with a protein extract from regenerating newt limb, suggesting that when stimulated by the appropriate factors, mammalian cells can form stem cells by dedifferentiation [21, 22]. Regeneration of the neural retina and lens in newts and *Xenopus* takes place by dedifferentiation and transdifferentiation of the pigmented retina, which can also regenerate itself [23, 24]. The pigmented epithelial cells of the retina thus can give rise to two other sets of cell lineages in addition to its own during regeneration of eye structures.

Resident stem cells, hepatocytes and dedifferentiated cells of amphibians can all be categorized as ***regeneration-competent cells*** (**RCCs**). Regardless of their position or function in the body, RCCs have two things in common – they are able to undergo changes in response to local maintenance or injury signals that allow them to proliferate and differentiate into new tissue and they can do this repeatedly, though the mechanisms involved may be different depending on the RCC.

Regeneration-competent cells have also been demonstrated in regeneration-deficient tissues and organs [25]. These stem cells are capable of differentiating into functional cells in vitro, indicating that their environment suppresses their regenerative capacity and/or favors the proliferation of scar-forming fibroblasts over that of the stem cells. This suggests that if large numbers of stem cells can be transplanted to the lesion, or if the injury environment in regeneration-deficient organs can be modified appropriately, regeneration of these organs may be possible. In this chapter, I review the activity of stem cells in two such regeneration-deficient organs, the central nervous system and the heart, and describe attempts to regenerate or rescue these tissues by stem cell transplants or by chemically inducing their regeneration from resident regeneration-competent cells.

## 2 Regeneration of Central Nervous System Tissues

### 2.1 Neurogenesis in the Adult Brain

Three decades ago, it was believed that the mammalian central nervous system generated no new neurons after birth, either to maintain neural populations or to regenerate new neurons in response to injury. However, labeling studies over the last three decades using $^3$H-thymidine, BrdU and dyes, have conclusively revealed the presence of neural stem cells (NSCs) in the striatum, hippocampus, and telencephalic lateral ventricles of the brain, where they migrate into the olfactory bulb and differentiate into neurons and glia [26–36]. In the presence of epidermal growth factor (EGF), these cells form neurospheres in vitro that react to antibodies to nestin, a cytoskeletal protein characteristic of embryonic neural precursors. Some cells of the neurospheres differentiate into neurons and glia upon withdrawal of EGF, as indicated by immunoreactivity to antigens specific for neural and glial cells, while other cells make new neurospheres, indicating that the NSCs can self-renew. Proliferating NSCs have been observed in other regions of the brain, but give rise only to new glial cells [37, 38].

The NSCs of the lateral ventricles and the hippocampus have been purified by FACS fractionation and tested for their ability to form neurons and glia under clonal conditions [36, 39]. A discrete nestin+ population of ependymal cells was isolated from the lateral ventricles that had low binding affinity for peanut agglutinin (PNA) and heat stable antigen (HSA), and did not express differentiated neuronal or glial cell markers. Neurospheres derived from this population differentiated into astrocytes (positive for glial acidic fibrillary protein, GFAP), oligodendrocytes (positive for the O4 protein), and neurons (positive for β-tubulin type III). Further evidence that the PNA$^{lo}$ HSA$^{lo}$ cells were NSCs was a sixfold reduction in the percentage of these cells in the mouse mutant *querkopf*, which has a greatly reduced number of olfactory neurons [36]. Human dentate gyrus NSCs purified by FACS and transfected with the green fluorescent protein (GFP) gene under the control of the nestin enhancer or the Tα1 tubulin promoter proliferated in vitro and differentiated morphologically, antigenically, and electrophysiologically as neurons [39].

Proliferating NSCs have been found in the hippocampus of mice and rats [37, 40–42], marmoset monkeys [32] and human patients given BrdU as part of a cancer study [33]. The hippocampus is crucial for the formation and maintenance of memories and for learning new information and tasks. Studies in vivo strongly suggest that the number of new neurons born in the hippocampus of mice is influenced by both physical and cognitive activity. Mice which ran on a treadmill, or were placed in an enriched environment, exhibited increased stem cell proliferation and neurogenesis in the dentate gyrus that was correlated with learning a maze faster than controls [43–45]. In mice and rats, newly born hippocampal neurons are integrated into the local circuitry and behave functionally as neurons, as shown by ultrastructural and immunochemical analysis of synapse formation and electrophysiological recordings [46]. The importance of generating new neurons to memory formation has been demonstrated by trace conditioning studies in rats. Rats conditioned to a noxious stimulus in the presence of the DNA methylating agent methylazoxymethane (MAM), which reduces the number of new neurons generated by 80%, showed a reduction of 50% in the frequency of conditioned responses [47].

Whether or not NSCs generate new neurons in the uninjured neocortex is controversial. BrdU labeling studies on macaque monkeys suggested that NSCs in the lateral ventricles of the prefrontal, posterior, parietal, and inferior temporal cortex migrate through the subcortical white matter into the cortex and differentiate into mature neurons [48]. These neurons were retrogradely filled by dyes injected into cortical areas known to be their projection targets, indicating that their axons had become part of the cortical circuitry. A similar study [49] also revealed ventricular NSCs that migrated to the cortex. However, detailed examination of sections with confocal optics suggested that the labeled cells were glia and that the close apposition of the labeled cells with neurons gave the illusion of differentiation into neurons. Another study [50] also found that BrdU-labeled ventricular cells entered the mouse cortex but differentiated as glia, remained undifferentiated, or died.

Injured mammalian brain tissue responds to injury by increased neurogenesis, but the response is minimal. Neurogenesis has been observed in the neocortex following the controlled destruction of a subset of pyramidal neurons in the lower layer (VI) of the cortex that project to the thalamus [50]. Subsequent proliferation of BrdU-labeled lateral ventricle cells was no greater than in uninjured animals, but cells labeled with BrdU migrated into the cortex and one to two percent of them expressed the mature neuron marker NeuN. Retrograde labeling indicated that these were pyramidal neurons that were functionally integrated into the normal circuitry of the brain.

Other studies have shown increased production of granule cells in the dentate gyrus of the hippocampus in rats and gerbils after neuronal degeneration induced by focal or global ischemia [51–54]. The increase is minimal however, replacing only a small fraction of the degenerated neurons, not enough for functional recovery. Intraventricular injection of FGF-2 and EGF boosted the number of regenerated neurons to 40% of the number lost [54]. The neurons were functionally integrated into the normal hippocampal circuitry as determined by microscopy, the electrophysiological properties of synapses, and the performance of the rats on behavioral tasks. Neonatal astroglia induce the differentiation of adult hippocampal NSCs into neurons in vitro, whereas the effect of adult astrocytes is only half that of neonatal cells [55]. Furthermore, the effect seems specific to hippocampal astrocytes, since spinal cord astrocytes do not support hippocampal neurogenesis. These results suggest that the low level of increase in neurogenesis in the injured hippocampus is due to deficient FGF-2 and EGF signaling from hippocampal astrocytes to NSCs.

### 2.2 The Spinal Cord Does not Regenerate, but Contains NSCs

After crush injury to, or transection of, the spinal cord, events are triggered that lead to the formation of a glial scar within the lesioned area [56–58]. Ischemia deprives the tissue of oxygen and glucose, and excess leakage of plasma from damaged blood vessels causes the cord to swell, killing many interneurons and glial cells outright. Toxicity due to overproduction of the neurotransmitter glutamate, free radical production by lipid peroxidation, and breakdown products of myelin and dead cells spread the initial damage to neighboring uninjured regions, causing many intact interneurons to undergo apoptosis. These interneurons are not replaced. Simultaneously, oligodendrocytes secrete individual myelin proteins (Nogo, myelin-associated glycoprotein, MAG) and other molecules (chondroitin sulfate proteoglycans, CSPGs) into the lesion that inhibit axon growth in vitro [59–61]. Lastly, an inflammatory response brings neutrophils and macrophages into the lesion to ingest bacteria and cellular debris, creating large cavities in the cord tissue, around which forms a glial scar. Axon regeneration is initiated in the injured spinal cord, but is blocked by the glial scar, which produces the same types of inhibitory CSPGs as oligodendrocytes [61].

The glial scar is formed by meningeal fibroblasts and NSCs residing in the ependymal layer and subventricular zone of the spinal cord. When the cord is injured, ependymal NSCs divide to form an amplifying population of stem cells in the subventricular zone, which subsequently differentiate as reactive astrocytes that participate in glial scar formation. However, when removed from the cord and cultured in vitro in the presence of EGF or FGF, these cells form neurospheres and differentiate into neurons and glia [34, 35, 62–66]. The importance of the glial scar as a physiological and mechanical barrier to preventing axons from crossing the lesion is highlighted by the fact that functional recovery of the transected mouse spinal cord can occur if the transection is made in such a way as to minimize injury to the dura, fibroblastic infiltration, and displacement of the cut ends of the cord, all of which minimize fibrosis [67]. Whether NSCs might also replace lost interneurons under these conditions is not known.

## 3 Regeneration of Cardiac Muscle

### 3.1 Skeletal Muscle Regenerates via Resident Stem Cells

Vertebrate skeletal muscle regenerates by the proliferation of stem cells called "satellite cells" located between the sarcolemma and the overlying basement membrane of the myofiber [68–70]. Satellite cells are themselves the product of a population of multipotent CD45+ muscle stem cells (MuSCs) that have both hematopoietic and myogenic developmental potential and originate in the bone marrow [71, 72]. CD45+ cells isolated from uninjured muscle of mice are unable to form muscle but can rescue the hematopoietic system when injected into irradiated mice [73]. However, CD45+ cells from injured muscle are able to regenerate muscle [74]. These observations suggest that under injury conditions, MuSCs replenish the supply of satellite cells used for muscle regeneration. The conversion of MuSCs to SCs involves Wnt signaling. Injured muscle produces several isoforms of Wnt proteins. MuSC proliferation and myogenic specification is significantly reduced by injection of the Wnt antagonists sFRP2 and 3 into regenerating muscle [75].

### 3.2 Cardiac Muscle Initiates Regeneration after Injury but Does Not Complete It

Significant increases in nuclear ploidy levels are observed in hypertrophied human and rat hearts subjected to prolonged functional overload or insufficiency [76–79], suggesting that cells divide to form new cardiomyocytes. The division of human heart cells after left ventricular infarction was confirmed in a study of human patients who died 4–12 days after infarction [80, 81]. Four percent ($2\times10^6$ cells) of the cardiomyocytes in the border zone of the infarcts were in mitosis, 84 times the value for control patients who had died of other causes and had no major risk factors for heart disease. This is clearly a regenerative response comparable to the initiation of axon sprouting in the injured spinal cord. However, as in the cord, the response is aborted and the infarcted region is infiltrated by fibroblasts that form a scar, compromising the pumping action of the ventricle. To compensate, the remaining myocardium undergoes compensatory hypertrophy.

What are these dividing cells? One possibility is that they are the progeny of stem cells in the heart that cycle at low frequency under normal conditions to maintain normal numbers of cardiomyocytes and respond to injury with more intense proliferation. To test this idea, Beltrami et al. [82] searched the uninjured rat myocardium for cells that are Lin-, but positive for the common stem cell markers c-Kit, Sca-1 and MDR-1. They detected such cells throughout the myocardium at a frequency of approximately one in every $1\times10^4$ cardiomyocytes. The cells were small, with a high nuclear/cytoplasmic ratio and were often found in clone-like clusters. Many of the cells in the clusters either expressed the mitotic marker Ki67 or were at early stages of cardiomyocyte differentiation. Cells with the phenotype [Lin–, c-Kit+, CD45–, CD34–] were isolated from the myocardium by FACS; these cells could differentiate in vitro into immature cardiomyocytes, smooth muscle cells and endothelial cells.

The cultured Lin–, c-Kit+ cells were able to rebuild the infarcted rat myocardium. When injected into the borders of the infarcts, they proliferated and regenerated fully mature myocardium, with resultant improvement in ventricular function. The numbers of newly differentiated cardiomyocytes were orders of magnitude greater than the number of cells injected and their DNA content was 2 N, indicating that the cells had proliferated without fusion with already differentiated cardiomyocytes.

A second possibility is that the Lin−, c-Kit+ cells do not reside in the heart, but are recruited from other tissues, such as the bone marrow. Quiani et al. [83] reported that up to 10% of the cardiomyocytes of female human hearts transplanted to male recipients were positive for the Y chromosome, suggesting that the dividing stem cells might be recruited from a host source. However, the presence of Lin−, c-Kit+ cells in the undamaged heart and the fact that no cells with their phenotype are detectable in the bone marrow, argues that they are resident to the heart [82].

A third possibility is that a small subset of cardiomyocytes undergoes dedifferentiation in vivo to take on the Lin−, c-Kit+ phenotype. Adult newt cardiac muscle regenerates from such a subset of cardiomyocytes [84]. Labeling and electron microscope studies [85, 86] suggest that newt cardiomyocytes undergo partial dedifferentiation, resulting in the disruption of compact myofibrils [87] and the disappearance of the intercalated discs [77]. The intercalated discs are the lines of juncture between adjacent cell membranes, and are the site of gap junctions that allow communication between cardiomyocytes. Loss of the intercalated discs allows single cells to be formed. The mechanism underlying newt cardiomyocyte dedifferentiation is not known, but might involve specific proteases produced within the cell, as well as remodeling of the extracellular matrix, as in newt limb regeneration [20]. Dedifferentiated cardiomyocytes in damaged adult newt heart synthesize DNA, divide, and redifferentiate into new cardiac muscle [88–90].

Mammalian cardiomyocytes can undergo partial dedifferentiation in vitro [91]. A high percentage of them (up to 25% ventricular and 63% atrial) synthesize DNA [76, 77, 92, 93], though only a very small percentage of (atrial) cells have been observed to divide. Entry into the cell cycle and dedifferentiation appear to be linked in cardiomyocytes. The E2F-1 transcription factor transfected into cultured ventricular cardiomyocytes induces DNA synthesis while simultaneously inhibiting the cardiac and skeletal α-actin promoters, sarcomeric actin synthesis, and serum response factor, which is crucial for the transcription of both sarcomeric α-actins in cardiac muscle [94]. The ability to partially dedifferentiate, which is not shared by mammalian skeletal muscle, may be related to a difference in the relationship between proliferation and differentiation in developing skeletal and cardiac muscle. Skeletal myoblasts first proliferate, then withdraw from the cell cycle and fuse to form multinucleated myotubes. Just prior to fusion the myoblasts express the transcription factors MyoD1 and myogenin [93]. These proteins function as a proliferation/differentiation switch that induces the expression of muscle-specific genes such as myosin and creatine kinase, while simultaneously inducing withdrawal of the myoblasts from the cell cycle. By contrast, the MyoD1 and myogenin genes are not expressed in cardiac myoblasts and cardiac myoblasts proliferate and differentiate simultaneously [93]. Thus cardiac myoblasts seem to have parallel programs for proliferation and cytodifferentiation, rather than serial programs linked by a single switch. This unusual situation is associated with the fact that the heart is one of the first organs to form during development and must become functional while still growing.

Why is the regeneration process not completed in the infarcted mammalian heart? Presumably, fibroblast proliferation and scar tissue formation outruns and perhaps suppresses the proliferation of the cardiac stem cells. The nature of this process might profitably be investigated in the MRL/MpJ mouse, which does complete the regenerative process in the infarcted heart [95]. Confocal microscopy demonstrated cells in which antibody staining for α-actinin and BrdU were colocalized, indicating that cells were proliferating and migrating into the infarct to regenerate cardiac muscle. These cells had a mitotic index of 10–20% as compared to 1–3% in controls. Revascularization was more evident at the injury site in the MRL mice. By day 60, the mutant hearts were virtually scar-free and echocardiographic measurements indicated a return of heart function to normal. The lack of scarring in the infarcted hearts of these mice provides a potential opportunity to analyze what is different about the injury environment in them vs the injury environment in normal mice.

# 4 Regenerative Medicine in the CNS and Heart

There has been some success in the experimental use of adult neural stem cells or precursors derived from ESCs to repair local or global neural deficits by cell transplants or by mobilizing endogenous NSCs. Most of the work (and the success) has been with diseases of the central nervous system and less with traumatic injury to the brain or spinal cord. Neural stem cells have a remarkable ability to migrate throughout the CNS, thus being potentially useful in treating global neurodegenerative diseases such as myelin deficiencies and Alzheimer's disease, as well as diseases in which neurodegeneration is more localized, such as Parkinson's or Huntington's [96].

## 4.1 Treatment of CNS Disorders

### 4.1.1 Demyelinating Disorders

Brustle et al. [97] injected glial precursor cells differentiated in vitro from mouse ESCs into the spinal cord and brain of rats that have the equivalent of a human demyelinating disorder, Pelizaeus-Merzbacher disease. This disease is caused by a mutation in the X-linked gene for myelin proteolipid protein (PLP). Two weeks after injection, electron microscopy and staining of the tissues with a probe to mouse satellite DNA and antibody to PLP showed that the injected cells had differentiated into oligodendrocytes that formed myelin sheaths around the axons.

Multiple sclerosis (MS) is an autoimmune CNS disorder in which myelin is damaged and astrocytes proliferate to form scar, resulting in the blockade of electrical impulses along nerve axons, loss of sensation and coordination, and in severe cases, paralysis and blindness [98]. Much work has been directed at halting the progression of the disease. Experimental work on a mouse model of MS, autoimmune encephalomyelitis (EAE), and clinical trials in humans suggest that attacks of MS can be aborted or diminished by blocking the $\alpha 4$ integrin molecule on the surface of the attacking immune cells, or by the use of agents that reduce the production of inflammatory cytokines and metalloproteases by inflammatory cells [99–101]. Regardless of how far the disease has progressed, when it is arrested we want to be able to repair whatever damage has occurred. Such repair has been accomplished in EAE mice by the intravenous injection of neurospheres derived from NSCs of the lateral ventricles of the brain [102]. The NSCs express the same $\alpha 4$ integrin expressed by attacking immune cells; they homed to sites of demyelination where they differentiated into oligodendrocytes and new neurons. Astrogliosis and axon damage were markedly reduced. Nearly 27% of the mice experienced remyelination with complete functional recovery from paralysis, whereas controls showed no sign of recovery. Interestingly, donor NSCs provided only 20% of the new oligodendrocyte precursors in sites of remyelination, the remainder being host cells. This observation suggests that the injected cells can regulate the behavior of host oligodendrocytes and astrocytes.

### 4.1.2 Parkinson's Disease

Parkinson's is an invariably progressive disease characterized by tremors at rest, akinesia and bradykinesia, muscle rigidity, postural instability and lack of facial expression [103]. Normal movement is regulated by the striatopallidothalamic output pathway (SPTOP). The neurotransmitter dopamine, produced by cells in the substantia nigra of the striatum, maintains a normal level of output from the SPTOP to the motor cortex [104]. Parkinson's is caused by the death of dopaminergic neurons (DANs) in the substantia nigra, leading to lower dopamine output, hyperactivity of the SPTOP and impaired motor function. The cause of neuronal death is unknown.

The three primary therapies for Parkinson's are (1) to increase the dopamine output of the remaining viable DANs by administration of L-dopa, which is taken up and converted to extra dopamine that compensates for the lost output of dying cells, (2) pallidotomy, and (3) inhibitory electrical stimulation of the subthalamic nucleus, which normally stimulates the globus pallidus [103, 104]. All these therapies reverse the akinesia, bradykinesia and rigidity of the disease, but not the tremors. L-dopa treatment produces severe side effects after several years and eventually has no effect on disease symptoms when the number of viable DANs becomes too low. Many symptoms recur a few years after pallidotomy or electrical stimulation of the subthalamic nucleus. None of the therapies slow the disease progression. Thus investigators have turned to NSC transplants and activation of local NSCs in the hope of a cure. These transplants have been performed on both human patients and a rat model of Parkinson's. In this model the neurotoxin β-hydroxydopamine is injected unilaterally into the striatum. The rats suffer muscle rigidity on one side and exhibit turning movements, as well as akinesia and bradykinesia, while the contralateral side serves as a control.

Fetal human mesencephalic cells (which include dopaminergic NSCs) were the first cells to be transplanted in an attempt to cure Parkinson's [103]. When injected into the striatum of Parkinson's patients, they differentiate into DANs and make synaptic connections with host neurons, restoring the activity of the SPTOP toward normal. The results of such transplants, however, have been highly variable. In the best cases there have been dramatic clinical improvements that have lasted five to ten years [103, 105]. The improvements are correlated with an increased output of dopamine, as visualized by increased uptake of $^{18}$fluoro-dopa in PET scans. In other cases, improvements have been minimal, or patients have continued to deteriorate. Autopsies of two patients who died, as well as transplant experiments on Parkinsonian rats indicate that this variation is due to differential survival of transplanted cells [103]. It is thought that a minimum of 80,000 DANs (~20% of the normal number of DANs in the human substantia nigra) are required to obtain a beneficial effect. Practical and ethical considerations, however, dictate that fetal tissues cannot be a reliable source of cells for transplant. Thus NSC transplants from other sources are desirable.

Transplantation of ESC derivatives, as well as adult NSCs have been explored as a treatment for Parkinson's.

Kim et al. [106] used a five-stage protocol [107] to differentiate mouse ESCs transfected with the nuclear receptor related-1 (nurr-1) gene into DANS. Nurr-1 is a transcription factor that promotes the differentiation of mesencephalic precursor cells into tyrosine hydroxylase (TH)+ dopaminergic cells in the presence of FGF-8 and Shh [108, 109]. Grafts of $5\times10^5$ TH+ cells were injected into the striatum of rats lesioned unilaterally by β-hydroxydopamine. The cells were functional DANS as assessed by morphological, neurochemical, electrophysiological and behavioral criteria. They released dopamine and extended axons into the host striatum. The axons formed functional synapses as indicated by significant recovery from amphetamine-induced rotation and improvement in step-adjusting, cylinder and paw-reaching tests. Since the brain is an immune-privileged site, the next step would be to determine whether DANs derived from human ESCs (or other allogeneic stem cells) would lead to improvement of motor function in the rat Parkinson's model.

Ourednik et al. [110] induced symptoms of Parkinson's in aged mice via systemic administration of 1-methyl-4-phenyl-1,2,3,6 tetrahydropyridine (MPTP) and implanted NSCs unilaterally into the substantia nigra one to four weeks after treatment. The cells migrated extensively into both hemispheres, where they reconstituted tyrosine hydroxylase (TH) and dopamine expression and function throughout the mesostriatal system. While some NSCs differentiated into TH+ cells, ~90% of the TH+ cells were host cells. This result strongly suggested that cell transplants can rescue host cells through their secretion of neurotrophic factors [111].

One such neuroprotective growth factor is glial-derived neurotrophic factor (GDNF). Kordower et al. [112] tested the effect of GDNF on symptoms of Parkinson's by injecting lentiviral constructs containing the GDNF gene into the substantia nigra of MPTP-treated monkeys. Three months later, motor performance of the monkeys was vastly improved and PET scans with $^{18}$fluorodopa showed an uptake of the tracer over 300% greater than in control animals injected with constructs containing the β-galactosidase gene. Histological studies revealed substantially more GDNF+, TH+ neurons in the substantia nigra and other parts of the striatum compared to controls. While it is possible that the transfected gene may have promoted the differentiation of endogenous striatal NSCs to DANS, it seems more likely that it rescued existing DANS, increasing the output of dopamine.

Cells of the subventricular zone proliferate in vitro in response to TGF-α, suggesting that this growth factor might be used to activate NSCs in the walls of the lateral ventricles of unilateral Parkinsonian rats that would then migrate to the degenerating substantia nigra and differentiate into DANs. To test this idea, a total of 50 μg of TGF-α was administered to the striatum via a shoulder-implanted minipump and cannula at the rate of 0.5 μl/h for two weeks [113]. The infused rats showed a 31.5% improvement in rotational behavior over controls. Studies of BrdU incorporation and immunostaining for nestin and differentiated neuron markers revealed that NSCs proliferated, migrated to the striatum and differentiated into new neurons, some of which were presumably DANs.

### 4.1.3 Huntington's Disease

Huntington's disease is a progressive disorder of movement accompanied by severe cognitive deterioration. The symptoms are associated with the death of multiple populations of striatal neurons, particularly medium spiny neurons. In contrast to Parkinson's disease, the cause of Huntington's is known to be a mutation in the N-terminal region of the huntingtin gene that results in a polyglutamine (CAG) repeat [114]. Huntington's can be mimicked in primates by injecting quinolinic acid into the striatum or by intramuscular injection of nitropropionic acid.

Grafts of fetal striatal tissue to the striatum of marmoset or macaque monkeys with induced Huntington's reversed the symptoms of the disease [115, 116]. Immunohistochemical studies indicated good survival and differentiation of the grafted neurons with establishment of functional connections with host tissue. In these non-mutant Huntington's models, some or all of the recovery might be due to a neuroprotective effect on compromised neurons by growth factors produced by the grafted cells. Evidence for such an effect has been obtained in experiments on monkeys [117]. The monkeys were first given intrastriatal implants, on one side, of polymer-encapsulated baby hamster kidney (BHK) fibroblasts transgenic for the human ciliary neurotrophic factor (CTNF). A week later, the implanted region was injected with quinolinic acid. The loss of striatal neurons was significantly attenuated in the CTNF implant region, compared to control implant regions with BHK cells lacking the CTNF construct.

Whether CNTF or other growth factors might have a protective effect on neurons in human Huntington's patients is unknown. However, grafts of fetal striatal tissue have been shown to survive in these patients and to alleviate symptoms of the disease [118]. Regardless of how much recovery is attributable to neuron replacement vs. rescue of host neurons, it is clear that the transplants have a positive effect.

### 4.1.4 Spinal Cord Injury

Several types of interventions have been used to prevent paralysis and promote regeneration of axons in the injured spinal cord of experimental animals, including the administration of neuroprotective agents (e.g., methylprednisolone, which is FDA approved for humans), antibodies to myelin proteins, and scaffolds to bridge lesions [25, 119].

Partial restoration of function after contusion of rat spinal cords has been accomplished by injecting neural/glial precursors differentiated from mouse ESCs in vitro into the lesion nine days after injury [120]. Staining with antibodies specific for mouse proteins and for glial and neuronal markers showed that many of the implanted cells survived, migrated throughout the injured area and differentiated into new interneurons, oligodendrocytes and astrocytes. These findings were positively correlated with regaining the ability to bear weight on their hind legs and by restoration of partly coordinated stepping movements.

The mechanism of recovery in these experiments was not clear. New mouse neurons may have made functional connections with host rat neurons, partly restoring signal transmission between brain and hind legs, or the mouse oligodendrocytes may have rebuilt myelin sheaths around demyelinated host axons, enabling them to conduct impulses again. The most likely explanation, however, is neuroprotection. Teng et al. [121] implanted NSCs in a PGA-based matrix into the hemisectioned adult rat spinal cord. There was significant axon regeneration, not from donor NSCs, but from host neurons, again demonstrating the capacity of NSCs to create an environment conducive to the salvation of endogenous cells.

## 4.2 Treatment of Myocardial Infarction

To establish "proof of concept" that grafted cardiomyocytes could integrate into host heart tissue, Soonpa et al. [122] injected suspensions of fetal cardiomyocytes from *lac-Z* transgenic mice into the uninjured ventricular myocardium of syngeneic host mice. The injected cells continued to proliferate and differentiated into mature cardiac muscle integrated with that of the host. Electron microscopic analysis indicated that the donor cells formed intercalated discs with the host myofibers, suggesting donor/host electrical coupling. No cardiac arrythmias were noted. Similar results were obtained after transplanting fetal cardiomyocytes into the ventricular myocardium of dystrophic dogs [123]. Cardiomyocytes differentiated from mouse ESCs in vitro were transplanted into the ventricular myocardium of *mdx* dystrophic mice [124, 125]. Staining with antibodies to dystrophin showed that the transplanted cells were stably integrated into the host cardiac muscle.

The finding that the rat heart contains regeneration-competent cells [82], or is able to create them by dedifferentiation, the ability of these cells to regenerate myocardium [82] and the presence of a proliferative response above baseline in infarcted human hearts [81], suggests that the human heart might contain such cells. Such cells might be isolated and expanded in vitro and high numbers of them transplanted into the injured myocardium to suppress scar formation and differentiate new heart muscle and blood vessels. The ability to do this with autogeneic cells would depend on the length of time it takes to expand the cells and whether a fibrotic response in the infarcted area could be delayed or reversed at a later time.

Another approach to the repair of cardiac muscle has been to inject skeletal muscle myoblasts or satellite cells into uninjured or injured heart tissue. C2C12 myoblasts injected into the uninjured myocardium of syngeneic mice withdrew from the cell cycle and formed typical myotubes [126]. No overt cardiac arrythmias were noted and the donor cells survived for as long as three months. Data were not obtained as to whether the skeletal myotubes became electrically coupled to host cardiomyocytes or whether their contractile activity contributed to the function of the myocardium. Taylor et al. [127] induced myocardial infarction in rabbit hearts by cryoinjury or coronary ligation. They injected $10^7$ skeletal myoblasts into the infarcts and found that many of them differentiated into cells with the characteristics of cardiomyocytes connected by intercalated discs, while others differentiated into multinucleated skeletal muscle. Cardiac function was improved, but whether the mechanism of improvement was due to contraction of the muscle formed by the donor cells or to a lessening of mechanical stiffness of the scar by the presence of new muscle, or both, was not clear. Others have reported that satellite cells transplanted into cryo-infarcted ventricular muscle of rats, rabbits and pigs integrated into the heart muscle, differentiated into cardiomyocytes and improved heart function [128–130].

The first phase I trial transplanting satellite cells into the damaged human heart was carried out by Menasche [131] on a 72-year-old patient suffering from severe congestive heart failure caused by extensive myocardial infarction. Satellite cells were isolated from a quadriceps muscle biopsy, expanded in vitro for two weeks and $800\times10^6$ cells (65% myoblasts) delivered into the myocardial scar via 30 injections with a small-gauge needle. Simultaneously, a double bypass was performed in viable but ischemic areas of the myocardium. Six months later, the patient's symptoms were dramatically improved. Echocardiogram showed evidence of new-onset contraction and fluoro-deoxyglucose PET scan showed increased metabolic activity of the infarct. The improvement was considered unlikely to be due to increased collateralization from the bypass region, because this region was far from the infarct. Since this trial, several other cardiac patients have been transplanted with satellite cells.

### 4.2.1 Transdifferentiation of Bone Marrow Cells in Regenerative Medicine

Under some experimental conditions, bone marrow stem cells (hematopoietic stem cells or mesenchymal stem cells) appear to have a developmental potential that exceeds their normal prospective fate. Differentiation of a cell into types other than its prospective significance or origin is commonly called transdifferentiation. The prospective potency of bone marrow stem cells has been tested by first labeling them in some way and then exposing them to signals they would otherwise never see. Cells can be labeled with transgenes for markers such as β-galactosidase or green fluorescent protein (GFP), by incorporation of lipophilic dyes, or bromodeoxyuridine (BrdU) incorporation. Natural markers such as the Y chromosome, or species-specific DNA sequences and antigens are also used. Assays for transdifferentiation include injecting the cells into irradiated or *scid* host mice (bone marrow reconstitution assay), injecting them into early embryos (chimeric embryo assay), or culturing them with other cell types in vitro. The presence of labeled cells having the morphology, molecular identity and function of differentiated cells from other lineages is considered evidence that the cells have a developmental potency wider than their prospective fate. Bone marrow cells have been found to possess remarkable developmental plasticity, raising the hope they could be transplanted into a variety of injured tissues and differentiate tissue that is in accord with their new location. Neural and cardiac cells are among the cell types that bone marrow cells are thought to be able to form.

### 4.2.2 Transdifferentiation to Neural Cells

Brazelton et al. [132] found that GFP-expressing bone marrow cells injected into lethally irradiated hosts not only reconstituted the hematopoietic system, but also transdifferentiated into neurons of the olfactory bulb, hippocampus, cortex and cerebellum. Confocal microscopy was used to show that GFP and neuronal-specific proteins (NeuN, NF-H) were expressed in the same cells and that these cells displayed neuronal morphology. An estimated 0.2–0.3% of the total number of neurons in the olfactory bulb were derived from bone marrow cells.

In a different assay, newborn female PU.1 null mice received intraperitoneal transplants of wild-type male bone marrow cells [133]. PU.1 is a transcription factor expressed in hematopoietic cell lineages. Mice lacking this factor lack immune cells and, without a bone marrow transplant, die within 48 h of birth. One to four months after receiving the male bone marrow, the female recipients were probed for Y+ cells that reacted to antibodies against NeuN and NSE. The results indicated that 2.3–4.6% of all cells in the brain were Y+, and that 0.3 –2.3% of cells were Y+, NeuN+ and NSE+. The latter cells were found in the hypothalamus, hippocampus, striatum, amygdala, and especially the cortex.

Adult rat and human MSCs were induced by β-mercaptoethanol or dimethyl sulfoxide/butylated hydroxyanisole treatment to differentiate at high frequency (up to 80%) in vitro into neural-like phenotypes expressing the neural markers NSE, NeuN, neurofilament-M, and tau [134]. Individual clones of cells were self-renewing, giving rise to both neurons and MSCs. In another set of experiments, human or mouse MSCs cultured in neural differentiation medium containing all-trans retinoic acid and BDNF differentiated at a low frequency into neural (0.5%) or glial (1%) phenotypes expressing NeuN or GFAP [135]. Human MSCs labeled with red or green fluorescent tracker dyes or mouse MSCs transgenic for *lac Z* were shown to express Neu-N at a frequency of 2–5% or GFAP at a frequency of 5–8% when co-cultured with mouse fetal midbrain cells. The morphology of these cells approximated that of immature neural and glial cells [135]. Experiments in vivo also suggest that MSCs can generate glial. When injected into the brain ventricles of neonatal mice, MSCs migrated throughout the forebrain and cerebellum, where they differentiated into astrocytes [136, 137].

Small cells (8–10 μm diameter) with a high nuclear/cytoplasmic ratio and the cell surface phenotype [MHC class I and II-, CD34-, CD44-, CD45-, c-Kit-, Thy-1lo, Sca-1lo, Flk-1lo] have been isolated from cultured bone marrow cells of Lac-Z+ (ROSA26) mice [138]. These cells, called multipotent adult progenitor cells (MAPCs), co-purify with MSCs and have several characteristics of embryonic stem cells, including LIF-dependence, high expression of telomerase, maintenance of telomere length for 100 population doublings, expression of SSEA-1 and the transcription factors Rex-1 and Oct-4. About one in $10^3$ cultured cells is capable of generating a clone of MAPCs. Chimeric embryo assay revealed that these cells can make contributions to a wide variety of tissues, including brain and myocardium.

### 4.2.3 Transdifferentiation to Myocardium

Bone marrow stem cells have been reported to transdifferentiate into cardiac cells in vivo and in vitro [139].

Orlic et al. [140] injected $1\times10^5$ Lin–, c-kit+ GFP-labeled HSCs from male mice into the myocardium of injured female hearts 3–5 h after infarction by coronary ligation. BrdU was administered each day for four to five days. Nine days after grafting, ventricular function was measured and sectioned hearts were immunostained for BrdU, GFP, Ki67 and cardiac-specific proteins. Hemodynamic function in the grafted animals improved by 40%. GFP+ cells incorporated BrdU, were positive for Ki67 and expressed cardiomyocyte-specific proteins. The grafted HSCs appeared to migrate into the infarct area and form new myocardium and blood vessels. Closely packed myocytes occupied 68% of the infarct and connexin 43 expression was detected at the surface of closely aligned cells, suggesting the formation of gap junctions.

In a similar study, ROSA26 (Lac-Z+) HSCs were used to reconstitute the marrow of lethally irradiated host mice, after which coronary ligation was performed [141]. Subsequent analysis revealed that, within the infarct, 0.02% of the cardiomyocytes, and up to 3.3% of the endothelial cells in small blood vessels, were donor-derived. However, little functional improvement was noted, perhaps because of the small number of new cells.

Kocher et al. [142] isolated hemangioblasts from human bone marrow, expanded them in vitro, and injected the cells into rat myocardial infarcts induced by coronary ligation. By two weeks, there was a significant increase in the microvascularity of the infarct, with human cells accounting for 20–25% of the increase. After 15 weeks, the reduction in hemodynamic function of the experimental hearts had decreased to 18–34%, compared to 48–59% in controls.

At least two MSC lines can differentiate as beating cells with a contractile protein profile of fetal ventricular cardiomyocytes, either spontaneously after long-term culture [143] or after 5-azacytidine treatment [144]. Both these lines express the cardiac-specific transcription factor Nkx2.5. Cultured rat MSCs labeled with BrdU were induced to differentiate as cardiomyocytes when transplanted into myocardium in vivo [145–147].

### 4.2.4 Is Transdifferentiation Real?

There are several alternatives to transdifferentiation that can explain the apparent developmental potency of bone marrow cells. First, a rare, primitive pluripotent stem cell, similar to the pluripotent ESC, might persist throughout development to reside in the bone marrow [148, 149]. This might be the cell that proliferates and differentiates into a wide variety of cell types when challenged in various assays. The existence of such a cell would be of great interest, since it would be a universal adult stem cell that could be used to regenerate any tissue. Some support for this idea comes from the bone marrow origin of multipotent CD45+ muscle stem cells (MuSCs) that have both hematopoietic and myogenic developmental potential [71, 72].

Second, pluripotent cells might be created by the dedifferentiation of adult bone marrow stem cells, reprogramming their nuclei so they are capable of responding to all sets of developmental signals. Acquisition of pluripotency in this way is a distinct possibility when cells are cultured for many generations in vitro, as was the case for the isolation of MAPCs [138].

Third, it is now apparent that fusion of donor bone marrow cells with differentiated host cells to form heterokaryons is a major phenomenon that can give the illusion of transdifferentiation. In these heterokaryons, the transcriptional profiles of the partner nucleus are adopted by the bone marrow nucleus through intracellular reprogramming, as opposed to responding to extracellular signals. Thus the apparent conversion of labeled mouse bone marrow cells into ESCs when co-cultured with ESCs was shown to be due to fusion of the bone marrow cells to the ESCs, at a frequency of $10^{-6}$ to $10^{-4}$. The labeled cells had a 4 N chromosome number and exhibited heterochromatins of both bone marrow cells and ESCs [150, 151]. Fusion has also been demonstrated in vivo. In an elegant experiment, ROSA26 transgenic mice with a floxed stop cassette fused to the Lac-Z gene were lethally irradiated and engrafted with whole bone marrow or HSCs from mice constitutively expressing Cre recombinase and GFP [152]. Fusion of donor cells with host cells could be detected because the Cre recombinase of the GFP-labeled cell excises the floxed stop cassette of the ROSA26 cell, allowing it to express Lac-Z. After ten months, β-galactosidase expressing cells with host morphology were observed at very low frequency in three tissues: brain, myocardium and liver. These cells were shown to have two nuclei with different morphologies. Weimann et al. [153] have shown in another study that bone marrow cells fuse with Purkinje neurons in the brain to form binucleate cells after injection into irradiated mice.

Other experiments are consistent with these results. Purified HSCs of ROSA26 mice transgenic for the LacZ gene failed to differentiate into neurons in irradiated C57B1/6 hosts when injected either before or after cortical stab injury [154]. A few β-gal positive cells were observed, but these were associated with blood vessels and did not have the morphology of neurons. Likewise, GFP-labeled HSCs injected into lethally irradiated recipients reconstituted the hematopoietic system, but only one GFP+ Purkinje neuron was found out of over 13 million cells examined in the brain. In addition, seven of 470,000 hepatocytes examined were GFP+, but no transdifferentiated cells were found in kidney, gut, skeletal muscle, cardiac muscle, or lung [155]. These low levels of GFP+ neurons could easily be explained by the fusion of donor HSCs to host neurons.

Since transdifferentiation of bone marrow cells in the chimeric embryo assay appears to occur at high frequency in some cases [138], it would be worth investigating whether donor/host cell fusion occurs at high frequency in this assay as well. Fusion at high frequency has already been demonstrated in the in vivo rescue of the liver of tyrosinemic (FAH-) mice by FACS purified HSCs transplanted at limiting dilution. Such transplants had appeared to give rise to hepatocytes that extensively reconstituted the FAH- liver [156]. When the study was repeated looking for evidence of cell fusion, these hepatocytes were found to be positive for molecular markers of both donor and host [157, 158]. Interestingly, donor cell markers were seen only after the host hematopoietic system had been reconstituted. This suggests that it is not the HSCs themselves that are fusing with host hepatocytes, but differentiated products of the HSCs such as macrophages, which fuse normally to form osteoclasts.

It would be interesting to know whether other adult stem cells fuse to host cells after transplantation, or whether this is a feature specific to bone marrow cells. There was no evidence for fusion of the Lin–, c-Kit+ cardiac stem cells to differentiated cardiomyocytes in the experiments of Beltrami et al. [82]. Cell fusion has been treated as an artifact of testing developmental potency, but Alvarez-Dolado et al. [152] have pointed out that it is also a normal developmental process. Multinucleated skeletal muscle is formed by the fusion of mononucleate myoblasts [159]. The multinucleated osteoclasts involved in bone remodeling arise by the fusion of monocytes [160, 161] and cardiomyocytes sometimes exhibit two nuclei under normal conditions [160]. Thus, fusion of endogenous or exogenous stem cells to surviving differentiated cells may actually be another mechanism by which injured tissue can regenerate itself.

## 5
## Summary and Conclusions

Regeneration-competent cells have been detected in the central nervous system and in cardiac muscle, where they appear to have a maintenance regeneration function. These cells increase their level of proliferation in response to injury, but the response is aborted and the repair process is dominated by fibrosis. This indicates that, although these tissues have an intrinsic ability to regenerate, the injury environment inhibits this ability. To overcome this inhibition, the injury environment needs to be either transplanted with regeneration-competent cells in sufficient numbers to provide their own proliferation and differentiation factors, or chemically modified to protect vulnerable endogenous differentiated cells and/or promote the proliferation of local regeneration-competent cells. Both these strategies have proven useful in attempting to establish a regenerative medicine for the CNS. The transplantation strategy has shown promise for a cardiac regenerative medicine.

It is likely that the regenerative medicine of the future will be a combination of these two approaches. Transplants of human adult stem cells or derivatives of human embryonic stem cells will be the first wave. For these to be successful, we will need to have a thorough understanding of the developmental signals required to direct the differentiation of ESCs to the desired cell type or to expand the numbers of adult stem cells to levels sufficient for transplant. Most importantly, we will need to know much more about the interactions of donor and host cells that integrate the transplanted cells into the host tissue architecture and establish the correct pattern of cell connections.

Will it be necessary, in the case of adult stem cells, to harvest them from the tissue that is to be repaired? This might be problematic for tissues such as those of the central nervous system and the heart. That is why the apparent ability of bone marrow cells to transdifferentiate has caused so much excitement. Bone marrow cells are relatively easy to harvest and expand, and if they can transdifferentiate into virtually any cell type when exposed to the right set of conditions, they could become a "universal stem cell". However, there is strong evidence that much of what has been considered to be transdifferentiation of bone marrow cells is actually attributable to fusion of these cells to host differentiated cells, particularly to CNS, cardiac, and liver cells. Chimeric embryo assays suggest that bone marrow cells can transdifferentiate at high frequency, but tests for cell fusion have not yet been thoroughly conducted in this assay. Nor do we know whether stem cells other than bone marrow exhibit fusion to host cells when transplanted. Fusion does not necessarily mean "artifact", because it is a normal developmental process in at least several tissues, and thus may be clinically useful as well. Whether the observed developmental potential of bone marrow cells has a single or multiple explanations matters little if the end result restores tissue structure and function.

If we are to transplant allogeneic adult stem cells or ESC derivatives to sites that are not immunoprivileged, we must contend with immunorejection. The preparation of donor-specific ESC lines by somatic nuclear transfer could eliminate the immunorejection problem, but this requires destroying a blastocyst specifically for one's personal needs, a process that is fraught with bioethical questions.

The ultimate in regenerative biology and medicine will be to understand what molecular attributes characterize a regeneration-competent cell of one kind or another, what signals are required to make such a cell proliferate and differentiate into site-specific tissue, and what environmental inhibitors might prevent this. With this kind of understanding, we would be able at the least to promote regeneration in regeneration-deficient tissue by the activation of resident regeneration-competent cells. If such cells did not exist at a nonregenerating site, we might be able to create them from local cells. Such a pharmaceutical approach to regeneration would be much simpler to use clinically than cell transplantation and would also be much less expensive.

The study of regeneration is enjoying a much-deserved renaissance. Significant understanding of regenerative mechanisms will be obtained not only with mammalian animal models, but with other vertebrate animals that are strong regenerators, such as amphibians [20]. Central nervous and cardiac tissues will be two of the primary targets for regenerative intervention. We may expect to see significant advances in treatments for injuries and diseases of these tissues within a decade.

## References


1. Morrison SJ, Shah N, Anderson DJ (1997) Cell 88:287
2. Fuchs E, Segre JA (2000) Cell 100:143
3. Ramos CA, Venezia TA, Camargo FA, Goodell MA (2003) Biotechniques 34:572
4. Weissman IL (2000) Cell 100:157
5. Watt FM, Hogan BLM (2000) Science 287:1427
6. Pain B, Clark ME, Shen M, Nakazawa H, Sakuri M, Samarut J, Etches R (1996) Development 122:2339
7. Brook FA, Gardner RL (1997) Proc Natl Acad Sci USA 94:5709
8. Hong Y, Winkler C, Schartl M (1998) Dev Genes Evol 208:595
9. Thomson JA, Itskovitz-Eldor J, Shapiro SS, Waknitz MA, Swiergiel JJ, Marshall VS, Jones JM (1998) Science 282:1145
10. Shamblott MJ, Axelman J, Wang S, Bugg EM, Littlefield JW, Donovan PJ, Blumenthal D, Huggins GR, Gearhart JD (1998) Proc Natl Acad Sci USA 95:13726
11. Smith A (2001 Embryonic stem cells. In: Marshak DR, Gardner R, Gottlieb D (eds) Stem cell biology. Cold Spring Harbor Laboratory Press, Cold Spring Harbor New York, p 205
12. Nichols J, Zevnik B, Anastassiadis K, Niwa H, Klewe-Nebenius D, Chambers I, Scholer H, Smith A (1998) Cell 95:379
13. Avilon AA, Nicolis, Pevny L, Perez L, Vivian N, Lovell-Badge R (2003) Genes Dev 17:126
14. Hanna LA, Foreman RK, Tarasenko IA, Kessler DS, Labosky PA (2003) Genes Dev 16:2650
15. Mitsui K, Tokuzawa Y, Itoh H, Segawa K, Murakami M, Takahashi K, Maruyama M, Maeda M, Yamanaka (2003) Cell 113:631
16. Chambers I, Colby D, Robertson M, Nichols J, Lee S, Tweedie S, Smith A (2003) Cell 113:643
17. Bucher NLR, Malt RA (1971) Regeneration of the liver and kidney. Little, Brown and Co, Boston
18. Michalopoulos GK, DeFrances MC (1997) Science 276:60
19. Brockes JP, Kumar A (2002) Nat Rev Mol Cell Biol 3:566
20. Stocum DL (2004) Amphibian regeneration and stem cells. In: Heber-Katz E (ed) Regeneration: stem cells and beyond. Springer, Berlin Heidelberg New York, p 1
21. Odelberg J, Kollhoff A, Keating MT (2000) Cell 103:1099
22. McGann C, Odelberg SJ, Keating MT (2001) Proc Natl Acad Sci USA 98:13699
23. Mitashov VI (1996) Int J Dev Biol 40:833
24. Raymond PA, Hitchcock PF (1997) Adv Neurol 72:171
25. Stocum DL (2004) Tissue restoration through regenerative biology and medicine. Springer, Berlin Heidelberg New York
26. Messier B, LeBlond CP, Smart I (1958) Exp Cell Res 14:224
27. Smart I (1961) J Comp Neurol 116:325
28. Altman J (1962) Science 135:1127
29. Altman J (1963) Anat Rec 145:573
30. Reynolds BA, Weiss S (1996) Dev Biol 175:1
31. Lois C, Alvarez-Buylla A (1994) Science 264:1145
32. Gould E, Tanapat P, McEwen BS, Flugge G, Fuchs E (1998) Proc Natl Acad Sci USA 95:3168
33. Eriksson PS, Perfilieva E, Bjork-Eriksson T, Alborn AM, Nordberg C, Peterson DA, Gage FH (1998) Nature Med 4:1313
34. Johansson C, Momma S, Clarke D, Risling M, Lendahl U, Frisen J (1999) Cell 96:25
35. Doetsch F, Caille I, Lim DA, Garcia-Verdugo JM, Alvarez-Bullya A (1999) Cell 97:703

36. Rietze RL, Valcanis H, Brooker G, Thomas T, Voss AK, Bartlett PF (2001) Nature 412:736
37. Gage FH (2000) Science 287:1433
38. Temple S (2001) Nature 414:112
39. Roy NS, Wang S, Jiang L, Kang J, Benraiss A, Harrison P, Restelli C, Fraser R, Couldwell WT, Kawaguchi A, Okano H, Nedergaard M, Goldman SA (2000) Nat Med 6:271
40. Rao MS (1999) Anat Rec 257:137
41. Momma S, Johansson SB, Frisen J (2000) Curr Opinion Neurobiol 10:45
42. Geuna S, Borrione P, Fornaro M, Giacobini-Robecchi MG (2001) Anat Rec 265:132
43. Kempermann G, Kuhn HG, Gage FH (1998) J Neurosci 18:3206
44. Van Praag H, Kempermann G, Gage FH (1999) Nature Neurosci 2:266
45. Van Praag H, Christie BR, Sejnowski TJ, Gage FH (1999) Proc Natl Acad Sci USA 96:13427
46. Van Praag H, Schinder AF, Christie BR, Toni N, Palmer T, Gage FH (2002) Nature 415:1030
47. Shors T, Miesegaes G, Beylin A, Zhao M, Rydel T, Gould E (2001) Nature 410:372
48. Gould E, Reeves AJ, Graziano MSA, Gross CG (1999) Science 286:548
49. Kornack D, Rakic P (2001) Science 294:2127
50. Magavi SS, Leavitt BR, Macklis JD (2000) Nature 405:951
51. Gould E, Tanapat P (1997) Neurosci 80:427
52. Liu J, Solway K, Messing RO, Sharp FR (1998) J Neurosci 18:7768
53. Jin K, Minami M, Lan JQ, Mao XO, Batteur S, Simon RP, Greenberg DA (2001) Proc Natl Acad Sci USA 98:4710
54. Nakatomi H, Kuriu T, Okabe S, Yamamoto S, Hatano O, Kawahara N, Tamura A, Kirino T, Nakafuku M (2002) Cell 110:429
55. Song H, Stevens CS, Gage FH (2002) Nature 417:39
56. Steward O, Schauwecker PE, Guth L, Zhang Z, Fujiki M, Inman D, Wrathall J, Kempermann G, Gage FH, Saatman KE, Raghupathi R, McIntosh T (1999) Exp Neurol 157:19
57. Fawcett JW, Asher RA (1999) Brain Res Bull 49:377
58. McDonald JW, Sadowsky C (2002) Lancet 359:417
59. Kapfhammer JP, Schwab ME (1992) Curr Opinion Cell Biol 4: 863
60. Niederost B, Zimmerman DR, Schwab ME, Bandtlow CE (1999) J Neurosci 19:8979
61. Filbin MT (2000) Curr Biol 10:R100
62. Weiss SM, Dunne C, Hewson J, Wohl C, Wheatley M, Peterson AC, Reynolds BR (1996) J Neurosci 16:7599
63. Chiang YH, Silani V, Zhou FC (1996) Cell Transpl 5:179
64. Kehl LJ, Fairbanks CA, Laughlin TM, Wilcox GL (1997) Science 276:586
65. Zhou FC, Chiang H (1998) Wound Rep Reg 6:337
66. Zhou FC, Kelley MR, Chiang YH, Young P (2000) Exp Neurol 164:200
67. Seitz A, Aglow E, Heber-Katz E (2002) J Neurosci Res 67:337
68. Mauro A (1961) J Biophys Biochem Cytol 9:493
69. Hinterberger TJ, Cameron JA (1990) Ontogenez 21:341
70. Carlson BM (2003) Dev Dynam 226:167
71. Seale P, Sabourin LA, Girgis-Gabardo A, Mansouri A, Gruss P, Rudnicki MA (2000) Cell 102:777
72. McKinney-Freeman SL, Jackson KA, Camargo FD, Ferrari, Mavilio F, Goodell MA (2002) Proc Natl Acad Sci USA 99:1341
73. Jackson KA, Mi T, Goodell MA (1999) Proc Natl Acad Sci USA 96:14482
74. Seale P, Rudnicki MA (2000) Dev Biol 218:115
75. Polesskaya A, Seale P, Rudnicki MA (2003) Cell 113:841
76. Oberpriller JO, Ferrans VJ, Carroll RJ (1983) J Mol Cell Cardiol 15:31

77. Rumyantsev PP (1991) Reproduction of cardiac myoctyes developing in vivo and its relationship to processes of differentiation. In: Rumyantsev PP (ed) Growth and hyperplasia of cardiac muscle cells. Harwood Press, New York, p 70
78. Borisov AB (1998) Cellular mechanisms of myocardial regeneration. In: Ferretti P, Geraudie J (eds) Cellular and molecular basis of regeneration. Wiley, New York, p 335
79. Borisov AB (1999) Wound Rep Reg 7:26
80. Beltrami AP, Urbanek K, Kajstura J, Yan S-M, Finato N, Bussani R, Nadal-Ginard B, Silvestri F, Leri A, Beltrami A, Anversa P (2001) New Eng J Med 344:1750
81. Anversa P, Nadal-Ginard B (2002) Nature 415:240
82. Beltrami AP, Barlucchi L, Torella D, Baker M, Limana F, Chimenti S, Kasahara H, Rota M, Musso E, Urbanek K, Leri A, Kajstura J, Nidal-Ginard B, Anversa P (2003) Cell 114:763
83. Quiani F, Urbanek K, Beltrami AP, Finato N, Beltrami CA, Nadal-Ginard B, Kajstura J, Leri A, Anversa P (2002) New Eng J Med 346:5
84. Bettencourt-Dias M, Mittnacht S, Brockes JP (2003) J Cell Sci 116:4001
85. Oberpriller JO, Oberpriller JC (1974) J Exp Zool 187:249
86. Bader D, Oberpriller JO (1978) J Morph 155:349
87. Nag AC, Healy CJ, Cheng M (1979) Tiss Cell 11:231
88. Oberpriller JO, Oberpriller JC (1971) J Cell Biol 49:560
89. Oberpriller JO, Oberpriller JC (1991) Cell division in adult newt cardiac myocytes. In: Oberpriller JO, Oberpriller JC, Mauro A (eds) The developmental and regenerative potential of cardiac muscle. Harwood, New York, p 293
90. Nag AC, Healy CJ, Cheng M (1979) Science 205:1281
91. Eppenberger ME, Hauser J, Baechi T, Schaub M, Brunner UT, Dechesne CA, Eppenberger HM (1988) Dev Biol 130:1
92. Claycomb WC (1991) Proliferative potential of the mammalian ventricular cardiac muscle cell. In: Oberpriller JO, Oberpriller JC, Mauro A (eds) The developmental and regenerative potential of cardiac muscle. Harwood, New York, p 351
93. Claycomb WC (1992) Trends Cardiovasc Med 2: 231
94. Kirshenbaum L, Abdellatif M, Chakraborty S, Schneider MD (1996) Dev Biol 179:402
95. Leferovich J, Bedelbaeva K, Samulewicz S, Zhang X-M, Zwas D, Lankford EB, Heber-Katz E (2001) Proc Natl Acad Sci USA 98:9830
96. Tate BA, Bower KA, Snyder EY (2001) Transplant therapy. In: Rao MS (ed) Stem cells and CNS development. Humana Press, Towota, p 291
97. Brustle O, Jones N, Learish R, Karram K, Choudhary K, Wiestler OD, Duncan ID, McKay RDG (1999) Science 285:754
98. Steinman L (1996) Cell 85:299
99. Yednock TA, Cannon C, Fritz LC, Sanchez-Madrid F, Steinman L, Karin N (1992) Nature 356:63
100. Steinman L (2001) Nature Immunol 2:762
101. Miller DH, Khan OA, Sheremat WA, Blumhardt LD, Rice GPA, Libonati MA, Willmer-Hulme A, Dalton CM, Miszkiel KA, O'Connor PW (2003) New Eng J Med 348:15
102. Pluchino S, Quattrini A, Brambilla E, Gritti A, Salani G, Dina G, Galli R, Del Carro U, Amadio S, Bergami A, Furlan R, Comi G, Vescovi A, Martino G (2003) Nature 422:688
103. Rosenthal A (1998) Neuron 20:169
104. Bjorklund A, Lindvall O (2000) Nature Neurosci 3:537
105. Lindvall O, Hagell P (2001) Clin Chem Lab Med 39:356
106. Kim J-H, Auerbach J, Rodriguez-Gomez JA, Velasco I, Gavin D, Lumelsky N, Lee S-H, Nguyen J, Sanchez-Pernaute R, Bankiewicz K, McKay R (2002) Nature 418:50
107. Lee SH, Lumelsky N, Auerbach JM, McKay RD (2000) Nature Biotechnol 18:675
108. Hynes M, Rosenthal A (1999) Curr Opinion Neurobiol 9:26

109. Wurst W, Bally-Cuif L (2001) Nature Rev Neurosci 2:99
110. Ourednik V, Ourednik J, Lynch WP, Snyder EY, Schachner M (2002) Nature Biotech 20:1103
111. Marconi MA, Park KI, Teng YD, Ourednik J, Ourednik V, Taylor RM, Marciniak AE, Daadi MM, Rose HL, Lavik EB, Langer R, Auguste KI, Lachyankar M, Freed CR, Redmond DE, Sidman RL, Snyder EY (2003) Neural stem cells. From in vivo to in vitro and back again. In: Sell S (ed) Stem cell handbook. Humana Press, Towota, p 191
112. Kordower JH, Emborg ME, Bloch J, Ma SY, Chu Y, Leventhal L, McBride J, Chen E-Y, Palfi S, Roitberg BZ, Brown WD, Holden JE, Pyzalski R, Taylor MD, Carvey P, Ling ZD, Trono D, Hantraye P, Deglon N, Aebischer P (2000) Science 290:767
113. Fallon J, Reid S, Kinyamy R, Opole I, Opole R, Baratta J, Korc M, Endo TL, Duong A, Nguyen G, Karkehabadhi M, Twardzik D, Loughlin S (2000) Proc Natl Acad Sci USA 97:14686
114. Huntington's Disease Collaborative Research Group (1993) Cell 72:971
115. Kendall AL, Rayment FD, Torres EM, Baker HF, Ridley RM, Dunnett SB (1998) Nature Med 4:727
116. Palfi S, Conde F, Riche D, Brouillet E, Dautry C, Mittoux V, Chibois A, Peschanski M, Hantraye P (1998) Nature Med 4:963
117. Emerich DE, Winn SR, Hantraye P, Peschanski M, Chen E-Y, Chu Y, McDermott PM, Baetge EE, Kordower JH (1997) Nature 386:395
118. Freeman TB, Cicchetti F, Hauser RA, Deacon TW, Li X-L, Hersch SM, Nauert GM, Sanberg PR, Kordower J, Saporta S, Isacson O (2000) Proc Natl Acad Sci USA 97:13877
119. Horner PJ, Gage FH (2000) Nature 407:963
120. McDonald JW, Liu XZ, Qu Y, Liu S, Mickey SK, Turetsky D, Gottlieb DL, Choi DW (1999) Nature Med 12:1410
121. Teng YD, Lavik EB, Qu X (2002) Proc Natl Acad Sci USA 99:3024
122. Soonpaa MH, Koh GY, Klug MG, Field LJ (1994) Science 264:696
123. Koh GY, Soonpaaa MH, Klug M, Pride HP, Cooper BJ, Zipes DP, Field LJ (1995) J Clin Invest 96:2034
124. Klug M, Soonpaa MH, Field LJ (1995) Am J Physiol 269:H1913
125. Klug M, Soonpa MH, Koh GY, Field LJ (1996) J Clin Invest 98:1
126. Koh GY, Klug MG, Soonpaa MH, Field LJ (1993) J Clin Invest 92:1548
127. Taylor DA, Atkins BZ, Hungspreugs P, Jones TR, Reedy MC, Hutcheson KA, Glower DD, Kraus WE (1998) Nature Med 4:929
128. Chiu RCJ, Zibaitis A, Kao RL (1995) Ann Thorac Surg 60:12
129. Kao RL, Chin TK, Ganote CE, Hossler FE, LI C, Browder W (2000) Cardiac Vasc Reg 1:31
130. Atkins BZ, Hueman MT, Meuchel JM, Cottman MJ, Hutcheson KA, Taylor DA (2000) Cardiac Vasc Reg 1:43
131. Menasche P (2002) Cardiac Vasc Reg 1:155
132. Brazelton TR, Rossi FMV, Keshet G, Blau HM (2000) Science 290:1775
133. Mezey E, Chandross KJ, Harta G, Maki RA, McKercher SR (2000) Science 290:1779
134. Woodbury D, Schwarz EJ, Prockop DJ, Black IB (2000) J Neurosci Res 61:364
135. Sanchez-Ramos J, Song S, Cardozo-Pelaez F, Hazzi C, Stedeford T, Willing A, Freeman TB, Saporta S, Janssen W, Patel N, Cooper DR, Sanberg PR (2000) Exp Neurol 164:247
136. Azizi SA, Stokes D, Augelli BJ, DiGirolamo C, Prockop DJ (1998) Proc Natl Acad Sci USA 95:3908
137. Kopen GC, Prockop DJ, Phinney DG (1999) Proc Natl Acad Sci USA 96:10711
138. Jiang Y, Jahagirdar BN, Reinhardt R, Schwarts RE, Keene CD, Ortiz-Gonzalez XR, Reyes M, Lenvik T, Lund T, Blackstad M, Du J, Aldrich S, Lisberg A, Low WC, Largaespada DA, Verfaille CM (2002) Nature 418:41

139. Jackson KA, Goodell MA (2003) Generation and stem cell repair of cardiac tissue. In: Sell S (ed) Stem cells handbook. Humana Press, Towota, p 259
140. Orlic D, Kajstura J, Chimenti S, Jakoniuk I, Anderson SM, Li B, Pickel J, McKay R, Nadal-Ginard B, Bodine M, Leri A, Anversa P (2001) Nature 410:701
141. Jackson KA, Majika SM, Wang H, Pocius J, Hartley CJ, Majesky MW, Entman ML, Michael LH, Hirschi K, Goodell MA (2001) J Clin Invest 107:1395
142. Kocher AA, Schuster MD, Szabolcs MJ, Takuma S, Burkhoff D, Wang J, Homma S, Edwards NM, Itescu S (2001) Nature Med 7:430
143. Jiang H, Hidaka K, Morisaki H, Morisaki T (2000) Cardiac Vasc Reg 1:274
144. Makino S, Fukuda K, Miyoshi S, Konishi F, Kodoma H, Pan J, Sano M, Takahashi T, Hori S, Abe H, Hata J, Umezawa A, Ogawa S (1999) J Clin Invest 103:697
145. Tomita S, Li RK, Meisel RD, Mickle DA, Kim EJ, Tomita S, Jia ZQ, Yau TM (1999) Circulation 100:II247
146. Wang S, Shum-Tim D, Galipeau J, Chedrawy E, Eliopoulos N, Chiu RJ-C (2000) J Thoracic Surg 120:999
147. Bittira B, Wang J-S, Shum-Tim D, Chiu RC-J (2000) Cardiac Vasc Reg 1:205
148. Owen M, Friedenstein AJ (1988) Ciba Found Symp 136:420
149. Caplan AI (1991) J Orthop Res 9:641–650
150. Terada N, Hamazaki T, Oka M, Hoki M, Mastalerz DM, Nakano Y, Meyer EM, Morel L, Petersen BE, Scott EW (2002) Nature 416:542
151. Ying Q-L, Nichols J, Evans EP, Smith AG (2002) Nature 416:545
152. Alvarez-Dolado M, Pardal R, Garcia-Verdiugo JM, Fike R, Lee HO, Pfeffer K, Lois C, Morrison SJ, Alvarez-Buylla A (2003) Nature 425:968
153. Weimann JM, Johansson CB, Trejo A, Blau HM (2003) Nature Cell Biol 5:959
154. Castro RF, Jackson KA, Goodell MA, Robertson CS, Liu H, Shine HD (2002) Science 297:1299
155. Wagers AJ, Sherwood RI, Christensen JL, Weissman IL (2002) Science 297:2256
156. Lagasse E, Connors H, Al-Dhalimy M, Teitsma M, Dohse M, Osborne L, Wang X, Finegold M, Weissman I, Grompe M (2000) Nature Med 6:1229
157. Vassilopoulos G, Wang P-R, Russell DW (2003) Nature 422:901
158. Wang X, Willenbring H, Akkari Y, Yorimaru Y, Foster M, Al-Dhalimy M, Lagasse E, Finegold M, Olson S, Grompe M (2003) Nature 422:897
159. Gilbert SF (2003) Developmental biology, 7th edn. Sinauer Associates, Sunderland
160. Ham AW, Cormack DH (1979) Histology, 8th edn. JB Lippincott, Philadelphia
161. Boyle WJ, Simonet WS, Lacey DL (2003) Nature 423:337

Received: January 2004

Adv Biochem Engin/Biotechnol (2005) 93: 161–172
DOI 10.1007/b99970


# Island Grafts: A Model for Studying Skin Regeneration in Isolation from other Processes

Dennis P. Orgill (✉)[1] · Charles E. Butler[2]

[1] Division of Plastic Surgery, Brigham and Women's Hospital, 75 Francis Street, Harvard Medical School, Boston, MA 02115, USA
*dorgill@partners.org*
[2] M.D. Anderson Cancer Center, University of Texas, Houston TX, USA



**Abstract** This chapter introduces the concept of an island graft, in which large full-thickness skin wounds are created with a small treatment area in the center. The island graft is a simple method for demonstrating the effect of epithelial derivatives on wound regeneration without using sophisticated biologic or genetic markers. In the absence of disease, mammalian wounds of the dermis close by scarring and contraction. The island graft permits observation of the test material within the wound without interference from the periphery for several days until wound contraction and epithelialization of the surrounding skin encroach upon the study area.

**Keywords** Island graft · Regeneration · Wound model

**List of Abbreviations**

| | |
|---|---|
| PTFE | Polytetrafluoroethylene |
| DRT | Dermis regeneration template |
| TCM | Tissue culture medium |
| FCS | Fetal calf serum |

## 1 Introduction

In this chapter we describe a methodology for studying skin regeneration isolated from other wound healing processes. Previous skin regeneration studies have been confounded by the difficulty of distinguishing regenerated tissue from adjacent intact host tissue or scar tissue. During the first few months following wound closure, regenerated tissue can be distinguished from mature tissue because it is less differentiated. As the regenerate matures, this distinction becomes less pronounced; the regenerate may be easily distinguished from mature host tissue, which has hair follicles, but not from scar tissue, which also lacks hair follicles.

Large defects in the integument are frequently seen in victims of thermal injury (burns) but are also common following trauma, tumor extirpation, and in chronic disease states. Without skin, the underlying tissues are susceptible to bacterial invasion and tissue desiccation, potentially leading to sepsis or tissue loss. In addition, long-term effects of these large wounds can result in significant scarring and contracture (Fig. 1).

Skin is composed of two layers: the epidermis and dermis. The epidermis provides a microscopic barrier to bacterial invasion and water loss. It is made up primarily of keratinocytes that divide, migrate upwards, undergo apoptosis, and form the protective layer *stratum corneum* that is continually renewed. Basal cells are located on the basement membrane and can repopulate the epidermis following a superficial injury. Skin adnexal structures such as hair follicles, sebaceous glands, and sweat glands invaginate into the dermis and are populated by basal cells. Following thermal or traumatic injury involving the entire epidermis and portions of the dermis, these cells can migrate, divide, and repopulate the epidermis. For injuries involving both the epidermis and the dermis, no basal cells or stem cells capable of epidermal regeneration are present.

Skin integrity can be restored only by central migration of keratinocytes from un-injured skin in the periphery (Fig. 2). This migration can often make it difficult to ascertain the origin of specific cells. Only in certain cases, where a disease process is observed in a large area of surrounding tissue, can one be reasonably certain about the circumstances leading to tissue changes. For example, the patient shown in Fig. 3 had bilateral radical mastectomies performed 30 years ago, followed by high does radiation therapy. The resection included skin, breast tissue, and the pectoralis major muscle. The patient developed a sarcoma in the central aspect of the treated area. Because the treatment area was much larger than the lesion found, we can reasonably infer that the subsequent lesion was related to the previous radiation therapy.

A variety of methods to close large skin defects have been proposed. Surgical advances have resulted in skin grafts that address the needs of different types of wounds. Most commonly, a split-thickness skin graft is taken from an uninjured area and fixed to the wound. In this procedure, a mechanical device

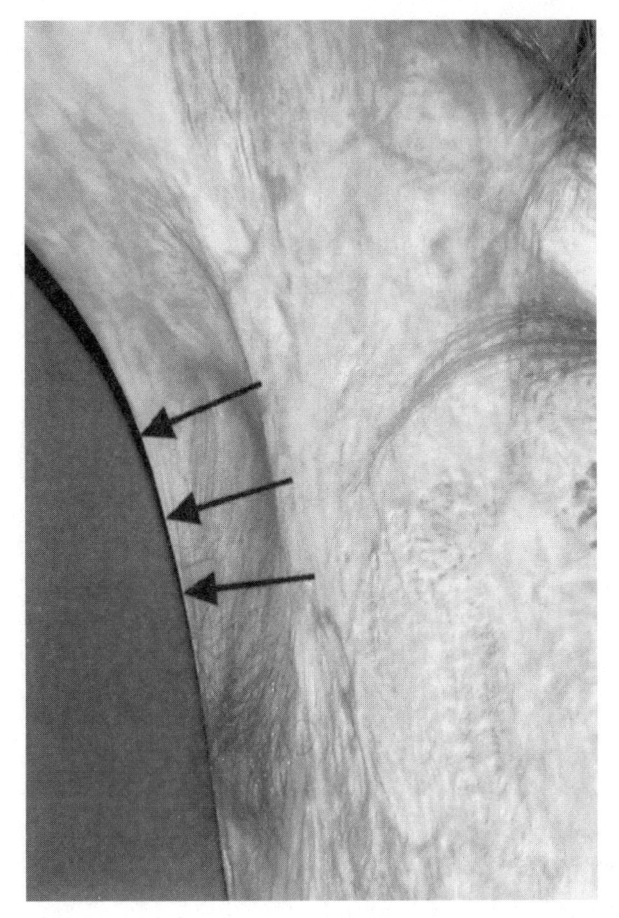

**Fig. 1** This patient had a large body surface area burn. Although treated aggressively with skin grafts, the patient formed two tight bands of scar tissue in the axilla (*arrows* point to scar tissue). (Figure courtesy of Brigham and Women's Hospital)

(dermatome) harvests skin of variable thickness (0.04 to 0.30 inches). These grafts include the entire epidermis and a variable thickness of the dermis. Because the resulting harvest site (donor site) wounds are partial thickness, the basal cells lining the skin adnexal glands have remained in place and re-populate the epidermis over a time period of 7 to 28 days to achieve wound closure. As the size of the donor site is often limited, a variety of strategies have been used to expand the donor skin including meshing (Fig. 4a, and Fig. 5), micrografts (Fig. 4b), and re-harvesting of the donor area after healing. Interestingly, many surgeons assume that thicker graft harvest sites heal at slower rates than more superficial sites. However, a recent study has shown similar healing rates for both due, perhaps, to the location of stem cells in the adnexal glands [1].

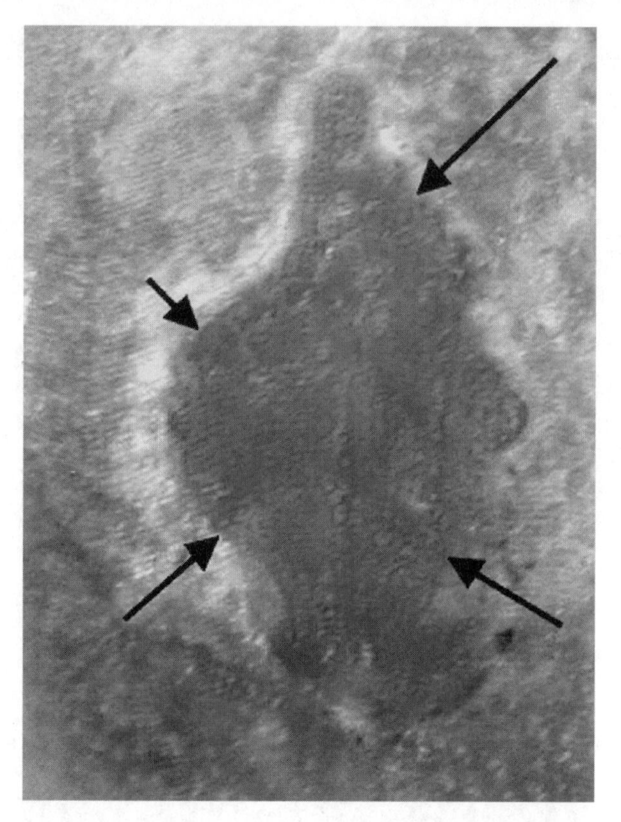

**Fig. 2** Wound healing by migration of the epidermis towards the center. (Figure courtesy of Brigham and Women's Hospital)

Meshed skin grafts are commonly made with a machine that cuts multiple parallel holes in the skin graft and allows it to expand in a direction perpendicular to the cuts. This produces a skin graft with multiple diamond-shaped openings 2–10 cm wide. Expansion ratios from 1.5:1 to 6:1 are commonly used today. The meshed procedure essentially converts a large wound into hundreds of much smaller wounds. The grafts adhere and keratinocytes migrate out from each of the edges to fill in the wounds. Although this can reduce the size of the donor site, the resulting meshed pattern scar on the graft site can be distressing to some patients.

Cadaveric allograft is often used as a dermal substitute. The epidermis of the allograft is rejected after several days but dermal elements incorporate into the wound and provide a good foundation for keratinocytes. More recently, a commercial process for making an acellular lyophilized dermal substitute has been successful. Yannas et al. have described in detail the design principles and methodology for synthesizing a porous biodegradable scaffold to induce regeneration of the dermis [4–7].

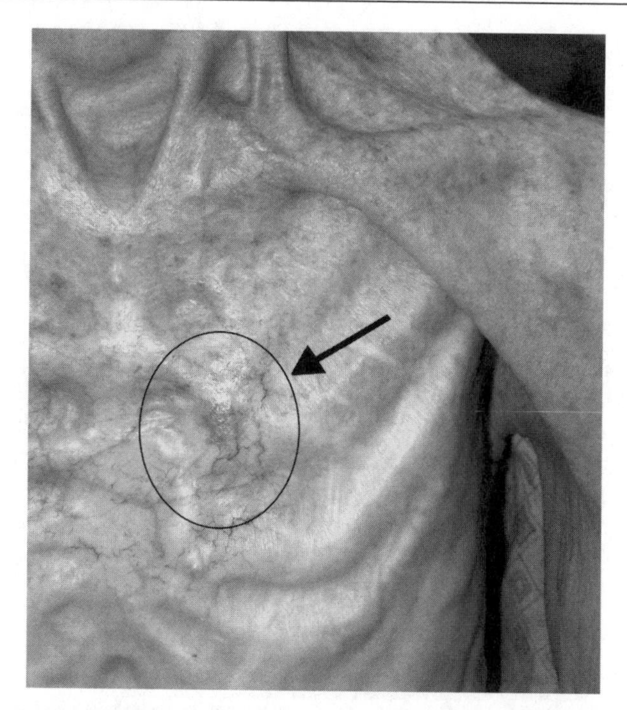

**Fig. 3** This elderly woman had a large treatment area by bilateral radical mastectomies followed by high dose radiation therapy. We can reasonably infer that the lesion in the center (sarcoma) is related to the radiation. (Figure courtesy of Brigham and Women's Hospital)

In contrast to meshed grafts, micro-grafts act by cutting the skin graft into very small pieces spaced throughout the wound. Skin grows out from each of these points and eventually forms a confluent sheet of epidermis. In both the meshed and micro-grafts, scarring can be problematic. To allow for the treatment of larger burns, and to help reduce scarring, tissue-engineering solutions have been sought. Pioneered by Rhinewald and Green [2], keratinocyte sheets are grown in cell culture from a very small donor area to produce cultured epithelial autografts (CEAs) that can be used to close wounds. CEAs were initially used directly on wounds but subsequent experience has shown that these materials work best when placed on dermis or a dermal substitute [3].

Island grafts have been previously demonstrated to separate wound edges from the test material [7]. Additional analyses of histological sections of regenerated tissue can be distinguished quantitatively from scar tissue by laser light scattering analysis [8]. In this method, scar tissue demonstrates a random light scattering pattern, whereas regenerated tissue or dermis appears as a more oriented pattern for regenerated.

A number of investigators have attempted to isolate the wound environment from the surrounding tissue through the use of wound chambers. Xu et al. used chambers on nude mice to graft skin substitutes [9]. Kangesu et al. used PTFE

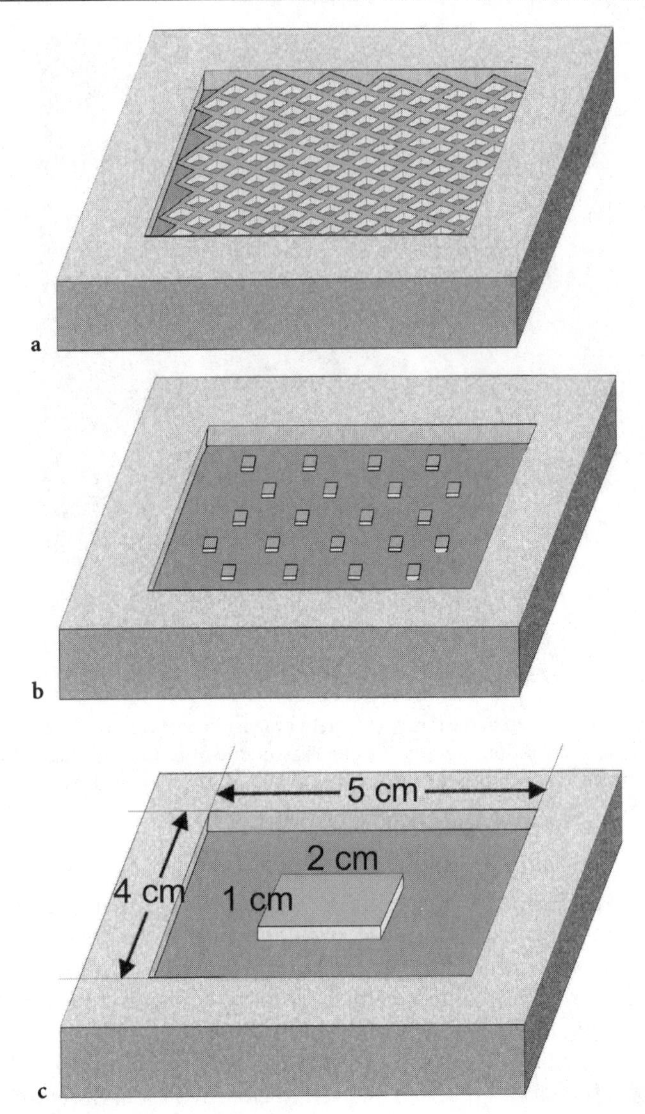


**Fig. 4** **a** Diagram of meshed skin autograft, which converts one large wound into multiple small wounds with migration of the epidermis from the mesh into the small intervening wounds. **b** Diagram of micrografts; the epidermis grows outward from the micrografts to coalesce and close the wound. **c** Diagram of island graft concept. The size of the graft is small relative to the size of the full-thickness wound surrounding it. This allows study of this portion of the wound in isolation before wound closure. (Figure courtesy of Brigham and Women's Hospital)

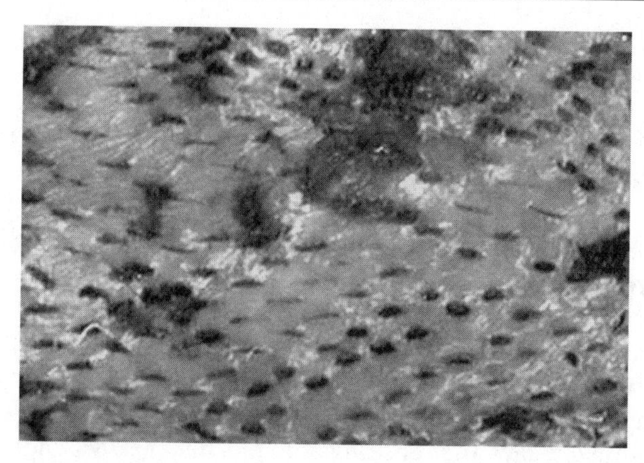

**Fig. 5** Meshed autograft. Meshed autografts reduce the donor site size but cause a corrugated appearance to the graft. (Figure courtesy of Brigham and Women's Hospital)

[polytetrafluoroethylene] chambers in a porcine model to study cultured keratinocytes [10]. The objective of these chamber models was to isolate processes of tissue synthesis from wound contraction and epithelial ingrowth. Both techniques have been criticized for introducing into the wound environment large foreign bodies, which may cause tissue reactions such as an inflammatory response.

We have developed a simple method to study the synthesis of new skin in relative isolation from surrounding mature skin (Fig. 4c). This method is related to the previously described technique of "open-style" grafting or "patch grafting" in which a fraction of the skin defect is left uncovered by the graft [11, 12]. Our procedure consists of grafting a small area in the center of a much larger, full thickness skin wound. This graft, based on a model of the extracellular matrix, has been reported to induce dermal regeneration [13]. If the distance between normal skin and the dermis regeneration template (DRT, a highly specific analog of keratinocyte-seeded extracellular matrix), is sufficiently large, neither epithelialization nor wound contraction interferes with the skin regeneration processes inside the graft. This model permits up to approximately 14–21 days of observation, depending upon specific characteristics of the wound and island graft.

The island graft procedure clearly separates graft healing from extraneous processes of the wound site, which encroach upon the regenerate and complicate its identification. These extraneous processes include: a) epithelialization from the wound edges over granulation tissue, b) contraction, or centripetal translation of the dermis and the epidermis of the wound edges, and c) scar synthesis which typically transforms the granulation tissue of the wound bed into a stiff highly aligned collagenous structure.

# 2 Materials and Methods

## 2.1 Copolymer Preparation

The bilayer graft was prepared as described by Yannas et al. [14]. The porous layer (in direct contact with the wound bed) was comprised of a graft copolymer of type I collagen and chondroitin 6-sulfate at a 92/8 weight ratio. The average molecular weight between cross-links was 12,500±5000 Daltons, and the average pore diameter was 70±30 μm. This insoluble macromolecular network loses approximately 50% of its dry weight to collagenolytic degradation processes within ten days after grafting onto full-thickness wounds in pigs [14]. The porous layer of the graft was covered with a 1 mm thick silicone film prepared by allowing the moisture sensitive prepolymer to cure at room temperature [4]. Keratinocytes were isolated and seeded into the copolymer as described below to create the DRT [14].

## 2.2 Keratinocyte Isolation and Seeding

Autologous split-thickness skin biopsies (ca. $0.02 \times 2.0 \times 3.0$ $cm^3$) were harvested from the backs of guinea pigs. Keratinocytes were isolated by incubating the biopsies in 0.25% phosphate buffered trypsin (GIBCO, Grand Island, NY), 1:250, pH 7.2, without calcium or magnesium, at 37 °C for 40 min, as described by Prunieras et al. [15]. The epidermis was removed from the underlying dermis using jeweler's forceps. The dermal layer of the graft was placed in a 50-ml polypropylene conical tube (Falcon 2098, Becton Dickenson, Oxnard, CA)and filled with 20–30 ml of tissue culture medium (TCM), consisting of Dulbecco's Modified Eagle's Medium (DMEM), 10% fetal calf serum (FCS), 100 U/ml penicillin and 100 mg/ml streptomycin [15]. The solution was vortexed (Vortex Genie, Model K-550-G, Scientific Products, McGraw Park, IL) at level 10 for 1 min. The dermal portion of the graft was removed and discarded after filtering the cellular suspension through sterile gauze. Cell yield was determined by using 0.4% Trypan blue (GIBCO, Grand Island, NY) exclusion and a hemocytometer. The cell suspension was centrifuged at 6500 g for 10 min and the volume of TCM adjusted to the desired density of viable cells.

Copolymer membranes were equilibrated in TCM for 30 min and then placed into specially constructed polycarbonate holders (Lexan, General Electric Co., Pittsfield, MA) with the silicone surface in contact with the holder. The cellular suspension was poured over the top of the porous membrane. The holder was placed into a standard four-inch centrifuge cup so that the centrifugal force vector was perpendicular to the plane of the membrane. The cell suspension was adjusted to the density levels of 0, $5 \times 10^5$, $1 \times 10^6$, and $3 \times 10^6$ cells/$cm^2$.

Temporal control was achieved by interrupting the experiment at day 14, before the contracting wound edges could reach the island graft.

## 2.3 Animal Model

NIH guidelines for the care and use of laboratory animals were observed. Full-thickness wounds were created in guinea pigs (n=3) as described by Orgill et al. [7]. White female Hartley guinea pigs (Charles River, Boston MA), 400–450 g, were caged individually and placed on a standard diet before surgery. A pre-operative dose of 0.1 mg/kg intramuscular tetracycline (Pfizer) was administered prior to anesthesia. Animals were anesthetized using a combination of nitrous oxide and halothane (Fluothane). Excision of the skin down to but not including the panniculus carnosus produced standard full-thickness 5.0×6.0 cm skin wounds. A 1.0×2.0 cm rectangle of DRT was sutured, porous side down, onto the center of the wound. The distance from the graft edge to the normal skin was approximately 2.0 cm at the time of grafting.

## 2.4 Wound Analysis

Biopsies that included adjacent tissue were taken at days 14 and 22, after the animal was euthanasized. Light microscopy specimens were stored in 10% buffered formalin, embedded in paraffin, sectioned, and stained with hematoxylin and eosin.

# 3 Results

The dimensions of the island graft and the wound on day 0 are illustrated in Fig. 5c. Contraction of wound edges proceeded at approximately the same rate when the wound bed was grafted with an island graft (n=3) as it did with ungrafted wounds.

On day 14 (Figs. 6, 7 and 8), the minimum separation between the graft perimeter and the wound edges was observed to be 12±2 mm. The 12-mm gap between the graft edge and the wound edge contained granulation tissue. Native epithelial cells were limited to the wound edge and had not approached the island graft. Biopsies revealed no evidence of scar formation within the graft and that fragments of the copolymer remained.

The thickness of the neoepidermis formed on day 14 shows a trend of increased thickness as the density of seeded keratinocytes increased. The hyperplastic nature of the neoepidermis that resulted from high cell seeding densities was temporary and was due in part to the presence of keratinocyte cysts that were extruded as the study progressed. The density of cysts increased

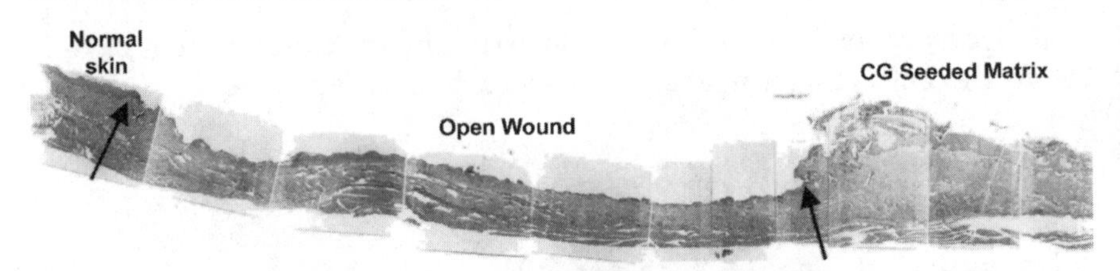


**Fig. 6** Composite histological cross section of the left half of a wound and island graft seeded with 500,000 cells/cm$^2$, day 14. Note the separation of normal skin from the CG-seeded matrix with a large open wound. (Figure courtesy of M.I.T.)

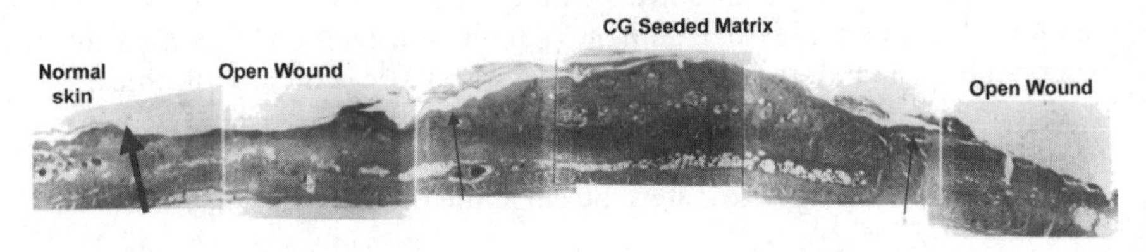


**Fig. 7** Composite histological cross section of island graft seeded with 1,000,000 cells/cm$^2$, day 14. The edge of the wound is shown (*thick arrow*) and the edges of the composite graft shown (*thin arrows*). *Between the thick and thin arrow* is open wound without epidermal coverage. Note the thickness of the CG-seeded matrix with multiple small cysts. (Figure courtesy of M.I.T.)

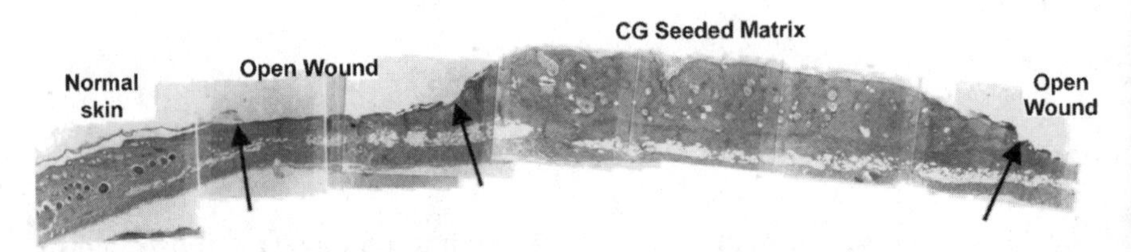


**Fig. 8** Composite histological cross section of island graft seeded with 3,000,000 cells/cm$^2$, day 14. Note that the CG-seeded matrix is surrounded by an open wound. Also note numerous cysts. (Figure courtesy of M.I.T.)

with the density of seeded cells. In an earlier fitted graft study [14], all cysts were shown to have been extruded by day 22, at which point, the thickness of the entire wound area grafted was near the physiological value. Cyst extrusion was not monitored in this study as all animals were sacrificed on day 14. The structure of skin synthesized from the island grafts was qualitatively similar to that which resulted from fitted grafting [13, 14, 16]. Absent seeded cells, epidermis was not synthesized simultaneously with the dermis, confirming results obtained from a previous guinea pig study [14] and two human studies [17, 18].

## 4 Discussion

This experimental design was based on a difference in dimensions between graft and wound bed. Spatial and temporal controls were used to achieve the desired separation between competing processes for wound closure. In summary, we observed that the processes leading to skin synthesis inside the graft could be clearly separated from other processes which lead to wound closure during healing. In particular, inspection of Fig. 7 suggests that epithelial cells migrating from the wound edges did not reach the perimeter of the island graft by day 14.

The objective of this chapter is to demonstrate a method for studying the process of skin regeneration separately from processes of wound healing. The skin regenerate from the island grafts appeared to be identical to fitted grafts as observed by Yannas et al. [13, 14, 16]. However, a detailed study of the similarities and differences between skin synthesized by island grafts and fitted grafts was not undertaken in this study.

The increase in neo-epidermal thickness on day 14 with increasing density of seeded keratinocytes is shown in Fig. 8. Subsequent biopsies in fitted grafts have shown that over time, an initially hyperplastic neoepidermis returns to a thickness which is near the physiological value, suggesting that early epidermal proliferation is a self-limiting process. This result is consistent with the presence of a mechanism for dimensional control of a regenerated organ.

We have presented a methodology to separate the results of physiological tissue synthesis from competing wound healing processes. This method is limited to grafts that are smaller than the wound area and to a timescale of a few weeks at most; beyond this point, wound contraction brings the wound edges into apposition with the edges of regenerated skin. However limited by time scale, the model is attractive because of its simplicity. The island graft can be used in conjunction with other methods, including genetic marking [19] and karyotyping [20], to study the long-term fate of seeded cells.

## References


1. Malpass KG, Snelling CF, Tron V (2003) J Plast Reconstr Surg 112:430
2. Rheinwald JG, Green H (1975) Cell 6:317
3. Krejci NC, Cuono CB, Langdon RC (1991) J Invest Dermatol 97:843
4. Yannas IV, Burke JF (1980) J Biomed Mater Res 14:65
5. Yannas IV, Burke JF, Gordon PL, Huang C, Rubenstein RH (1980) J Biomed Mater Res 14:65
6. Dagalakis N, Flink J, Stasikelis P, Burke JF, Yannas IV (1980) J Biomed Mater Res 14:511
7. Orgill DP, Yannas IV (1998) J Biomed Mater Res 39:531
8. Ferdman AG, Yannas IV (1993) J Invest Derm 100:710
9. Xu W, Germain L, Groulet F, Auger FA (1996) J Burn Care Rehab 17:7
10. Kangesu T, Navasaria S, Manek HA, Shurey CB, Jones CR, Fryer PR, Leigh IM, Green CJ (1993) Br J Plast Surg 46:393–400

11. Billingham RE, Medawar PH (1955) J Exp Biol 28:385
12. Billingham RE, Medawar PH (1955) J Anat 89:114
13. Yannas IV, Burke JF, Orgill DP, Skrabut EM (1982) Science 215:174
14. Yannas IV, Lee E, Orgill DP, Skrabut EM, Murphy GF (1989) Proc Natl Acad Sci USA 86:933
15. Prunieras M, Delecluse C, Regnier M (1980) Phar Ther 9:271
16. Yannas IV, Burke JF, Warpehoski M, Stasikelis P, Skrabut EM, Orgill DP, Giard DJ (1981) Trans Am Soc Artif Intern Organs 27:19
17. Lyerly HK, DiMario JM (1993) Arch Surg 128:1197
18. Burke JF, Yannas IV, Quinby WC Jr, Bondoe CC, Jung WK (1981) Ann Surg 194:413
19. Heimbach A, Luterman, Burke JF, Cram A, Herndon D, Hunt J, Jordan M, McManus W, Solem L, Warden G, Zawacki B (1988) Ann Surg 208:313
20. Hull BE, Sher SE, Rosen S, Church D, Bell E (1983) J Invest Dermatol 81:436

Received: February 2004

# Author Index Volumes 51–93

*Author Index Volumes 1–50 see Volume 50*

# Subject Index

Printing: Krips bv, Meppel
Binding: Litges & Dopf, Heppenheim